ENERGY AND ENVIRONMENT

ENERGY AND ENVIRONMENT

PETER E. HODGSON

Bowerdean Briefings
SERIES EDITOR: PETER COLLINS

BOWERDEAN
Publishing Company Limited

First published in 1997 by the Bowerdean Publishing Company Ltd.
of 8 Abbotstone Road, London SW15 1QR

British Library Cataloguing-in-Publication Data.
A catalogue record for this book is available from the British Library.

ISBN 0 906097 74 6

Designed by the Senate

Printed in Malta by Interprint Limited

CONTENTS

INTRODUCTION

All over the world people are realising more and more clearly that the increase in population and the growth of industries are seriously threatening the environment. The more people there are, the more energy will be needed, and this means more factories, more power stations, and therefore more pollution. We are gradually poisoning the earth, and if it dies, then we all die. We just cannot go on like this. There are limits to the growth of population and living standards, and if these are forgotten our planet will be poisoned forever.

The present population of the world is about five billion people, and this will rise to about six billion by the year 2000. On the average, it is doubling about every 35 years. People expect higher living standards and, as these depend on energy, the demand for energy is rising even more rapidly.

These are average figures; the reality is more complicated. In some regions, such as Europe and North America, the rate of population growth is small, whereas in Africa and Asia it is much more rapid. There are also large differences in living standards and energy consumption is very unevenly spread among the peoples of the world. The average energy consumption per person for the whole world is about 2 kWyr/yr, but about 70 per cent of the world's population have to try to survive with much less than this. At the other extreme, people in the developed countries use around 10 kWyr/yr. Even if we adopt the relatively modest goal of increasing the average energy consumption to 3 kWyr/yr over the next century, we would have to triple the world 's energy production, taking into account the increase in population over the same period. The disparity between energy consumption, and thus living standards, between the richer and the poorer countries is a measure of the ultimate energy need. Can this be achieved without destroying the environment?

Where is all this energy to come from, and can we ensure that it is generated without polluting the environment? There are many possible energy sources, and the familiar ones, such as coal and oil, are discussed in Chapter 1. There are also the renewable energies and nuclear power, and these are discussed in Chapter 2.

To decide the best energy policy, all these possible sources must be evaluated using five criteria, namely capacity, cost, safety, reliability and effects on the environment. This must be done separately for each country, as they differ greatly in their needs and natural resources. To ensure objectivity, the data for this discussion should be expressed in numerical form as far as possible. This can be done quite easily for the capacity, cost and reliability, and to some extent for safety, but is very difficult when we come to consider the effects on the environment.

The environment is important for our health and this is discussed in Chapter 3. All energy sources have more or less destructive effects on the landscape, and these are described and compared in Chapter 4. The effects of industrial waste are considered in Chapter 5.

The sensationalism and distortions of the mass media make it very difficult for people to reach a balanced view on energy questions and for Governments to implement sensible and realistic energy policies. These matters concerning the mass media and politics are discussed in Chapter 6.

The most difficult problems are those of the people in the developing countries. These people desperately need energy to raise their standard of living above starvation level in many cases. How can they afford to do this? If they do, what will be the effect on the environment? These questions are discussed in Chapter 7.

Among the large-scale sources is nuclear power, but this is linked to nuclear weapons. The resulting problems for world security are considered in Chapter 8.

If we are to survive, we need to take care of the earth, so that it can continue to provide what we need. Inevitably, each generation uses up some of the earth's resources, and this needs to be done with care and moderation. Still more importantly, the earth must be kept clean and unpolluted. Our responsibility for the earth is discussed in Chapter 9.

Although these subjects are discussed separately for convenience, they are closely interlinked. The use of one type of fuel depends on the availability of others. The costs determine its economic viability, and hence the possibility of minimising any adverse effects on the environment. The choice of energy source depends on balancing competing criteria that are not always commensurable. Increasing safety is generally costly and in extreme cases the cost of increasing safety to an acceptable level may be so high that another source has to be found. Likewise, it is expensive to reduce the effects on the environment and what is desirable is to some extent subjective. Furthermore, in a democratic society there are many other worthy projects competing for the available resources.

Another area of uncertainty concerns the data on which the decisions are based, and the difficulty of extrapolating into the future. Very frequently the data are inadequate, and it is not clear how best to estimate future trends. It is not, however, possible to wait until more data are available because the problems are urgent and the decisions must be taken now.

In addition to all this, there is always the possibility of completely new scientific and technical developments that either reveal a new energy source, or alter the potentialities of an existing one. Thus, nuclear power is a new source that was not seriously considered until the 1940s, and new devices such as improved photocells or superconducting transmission lines can alter the practicability and economics of existing sources.

Another consideration is the tension between short-term and long-term energy policies. Politicians seldom think beyond the next election, but concentrate on immediate problems and short-term solutions. Energy problems have a much longer timescale and so the decisions that would ensure a reliable energy supply for the next generation are seldom made.

It is thus not possible to give clear-cut answers to the many problems connected with energy and the environment. The aims of the present study are more modest, namely to provide some of the factual information on which an informed judgement must be based, and thus to show where is the area of fruitful discussion. All too often the inherent difficulty of the subject, the prevalence of politically driven propaganda campaigns and the sensationalism of the mass media implants in the minds of the general public ideas that are contrary to all objective evidence. This not only makes much of the contemporary debate completely unrealistic but inevitably influences Government policy, leading to decisions that are far from ideal and, in some cases, highly dangerous. This is a serious defect of our modern democratic society.

A particularly pernicious idea is that no energy source is acceptable unless it is shown to be perfectly safe. Assuming this, it is easy to build an impressive case against any energy source, complete with horror stories of accidents. Unfortunately, no energy source is perfectly safe; all exact their toll of deaths and injuries to the workers in the industry and the general public as well as harming the environment. Since none of this by itself can rule out any particular source, we then have to evaluate the hazards of each source in turn, and use the results to make the best practicable decision.

The most controversial subject concerning energy and environment is that of nuclear power: can it provide safe and limitless energy, or is it dangerous, costly and earth-polluting? Feelings run high concerning the safety of nuclear reactors and the disposal of radioactive waste, and so these questions are discussed in detail. Other equally important but much less controversial questions are given less attention. Popular beliefs, such as the desirability of what are called the benign renewables, are critically examined. Many important questions, particularly concerning international politics, are only briefly sketched. Others, such as population policy and the extent of our responsibilities to other countries and to future generations are vital moral questions that are beyond the scope of the present work. Nevertheless, it is hoped that the discussion will suffice to show the complexity and vital importance of thinking about our energy supplies and their effects on the environment.

I

THE ENERGY CRISIS

1.1 The Need for Energy

We need energy to cook our food, to light and heat our homes, and to drive our transport. We cook our food by burning wood, coal, gas or oil, or by electricity. We light and heat our homes with the same energy sources. We need petrol and oil for our cars, and coal, oil or electricity for our trains, ships and aeroplanes. The factories that make our clothes, our furniture, our refrigerators, our dishwashers, our packaged food all depend on energy.

Without energy all this would stop. Nothing could be made in factories. Cars, trains, ships and aeroplanes would no longer be possible. Our homes would be dark and cold, and we would have to eat raw food. Our lives would be hardly worth living; indeed, we would not survive for very long.

The population of the world is increasing, and so are the energy demands of each person. Each of us wants more and more energy. We want a larger car or a bigger house. We want more clothes and more labour-saving devices. We want to go further away for our holidays. If these demands are met, then each year we use up more energy than we did the previous year.

The result is that the amount of energy being used in the world is increasing even faster than the population. Actual consumption of energy does not, however, provide a true measure of world energy needs, because many people desperately need energy but do not get it. People in Europe, North America and Japan have a high standard of living and on the average use more than ten times as much energy per person as people in the poorer countries of Asia and Africa. If we were to give all the people in the poorer countries as much energy as those in Europe, North America and Japan we would have to multiply the world energy production at least tenfold. This would not only be economically very difficult, and would take many years, but we would have to consider very seriously the effects on the environment.

The energy needs of the rich and the poor are entirely different. For the rich, more energy means a second car, more travel, more labour-saving devices and so on. For the poor it is literally a matter of life or death. Most of the world's poor lack the energy for the basic necessities. For them, energy shortage means hunger instead of food, shacks instead of houses, cold instead of warmth, disease instead of health; in short, poverty instead of affluence.

This huge difference raises urgent questions. Is it right that such a gap should exist? What can we do about it? Can we ignore the poor and continue to live in luxury? Should we not reduce our energy consumption to make more available to the poor? Even if we were to do this, it would only be a small contribution to solving the problem. Whichever way we look at it, the world needs more energy.

1.2 Energy Sources

Energy comes in many forms. We can get energy in the form of heat by burning wood, coal and oil. We can use this heat to boil water and make the steam drive a turbine to generate electricity. This is more convenient to use because it is clean and convenient, and can easily be sent from one place to another. The turbine can also be driven by falling water: this is hydroelectric power. We can also get energy from the wind, in some places from hot springs, and possibly from the tides and the ocean waves. More recently we have learned how to obtain energy from the nucleus of the atom.

Which source of energy we use depends on a number of considerations such as availability, cost and reliability, and has to be decided for each region. We can get it more easily, cheaply and safely in some ways than in others. We therefore have to consider how much energy we can get from each source, and its cost, both direct and indirect. The direct costs are those of mining, extraction of the fuel from the ground, transport to the power station, construction and maintenance of the power station, and transmission of the energy to the consumer. The indirect costs include pollution and other effects on the environment, dangers to the health of the workers and to the whole population. To decide these questions we have to consider each energy source in turn.

1.3 Wood

Over the centuries, we have found increasingly concentrated and convenient sources of energy to satisfy our needs. From earliest times until the Middle Ages we relied mainly on natural fuels such as wood for heating, on animal and vegetable oils for lighting and on primitive windmills and waterwheels for grinding corn. The demand for wood soon exceeded the supply. Many forests bordering the Mediterranean were cut down in ancient times and those of central Europe followed in the Middle Ages.

The effects of energy demands on the environment are as old as history. In ancient times the countries along the southern shores of the Mediterranean supported large populations. Over the years the forests were cut down, the soil eroded and the once-fertile countryside reduced to a desert. Forests of willows, birches and pines thrived in the Outer Hebrides until they were burnt during the Viking invasions in the ninth and tenth centuries. They never recovered.

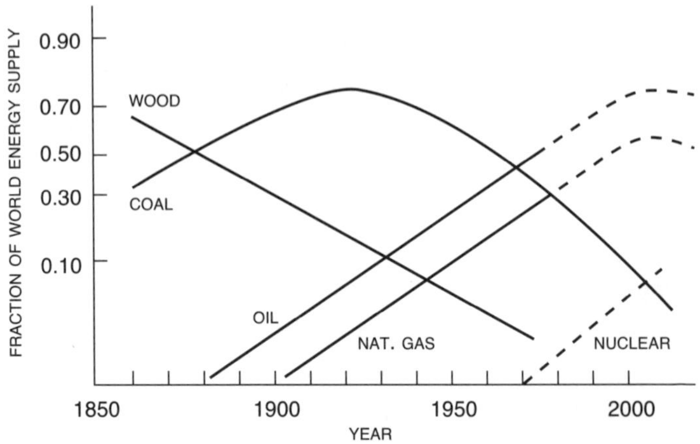

FIGURE 1. *The fraction of world energy supplied by different sources, showing how coal replaced wood, and then how oil and gas replaced coal. What will replace the oil and the gas when they eventually run out? (Hafele 1981)*

Wood is still used extensively as fuel, together with other organic material that would be better returned to the soil. There are proposals to establish plantations of fast-growing plants (collectively known as biomass) that could be harvested and burnt as fuel. Unfortunately, this would use up large areas of agricultural land, and involves serious problems of handling and transportation, so it is impracticable as a large-scale contributor to world energy needs.

1.4. Coal

As soon as coal was found, it was used instead of wood. Increasing quantities were mined in many European countries, and provided most of the power behind the industrial revolution in the nineteenth century. Before the development of railways, it was inconvenient to transport in large quantities from one place to another, and so the industries and people moved to the coalfields. The huge increase in the population of Europe in the nineteenth century, the development of manufacturing industries, railways and steamships, were all made possible by coal. There are still very large deposits of coal in many countries, enough for several hundred years at the present rate of consumption.

Coal will certainly remain a major source of energy for the foreseeable future.The technology is well understood and it is a familiar and accepted source of energy. There are, however, several disadvantages of coal that deserve serious consideration. Coal mining is dangerous, dirty and unpleasant, and increased coal production may mean more people working underground, and this must not be accepted lightly. In addition, as described in Chapter 5, coal burning on a large scale causes serious pollution.

1.5 Hydroelectric power

Electricity can also be generated by rivers in mountainous countries. To ensure a continuous supply of water, the rivers are dammed to form large lakes, and the water allowed to flow steadily through the turbines. Hydroelectric power stations provide much of the energy in countries such as Switzerland and Norway, which have many fast-flowing rivers. Most of the suitable rivers in the developed countries have already been used, but there are still many opportunities for the further development of hydroelectric power in developing countries, especially in South America, Africa and Asia. Even if all these are developed, however, it has been estimated that hydroelectric power can never supply more that about 10 per cent of the world's energy needs. Thus, although it is very important, and deserves to be further developed in suitable places, hydroelectric power is not the way to solve the energy crisis.

Unlike coal and oil, hydroelectric power has the advantage that it is inexhaustible; it does not use up the resources of the earth. It is, however, very destructive of the environment.

1.6 Oil

In some places oil oozes out of the ground, and this has been known since ancient times. In began to be used in large quantities when oil wells were drilled. The first was in Pennsylvania in 1859 and thereafter production rose rapidly. Oil has a higher energy content and is much easier to extract from the ground than coal, and to transport from the well to the consumer. It was therefore not necessary for the population and the industries to move to the oil wells. Instead, huge tankers carry the oil across the oceans of the world, from the Middle East and Alaska to Europe, the United States and Japan.

As the present century progressed, the use of oil rose rapidly, and in many industries it took the place of coal. At the same time, it made possible a whole new range of industries, such as plastics, drugs, paints dyes and so on, as well as air and motor transport.

Natural gas is often found associated with oil, and also independently. It can be used for heating and lighting, but in the early days there was no practicable way of transporting it to where it could be used, and so it was often wasted. Subsequently, gas was made from coal and used for street lighting, home heating and lighting. It has now been replaced with lighting by electricity, which is cheaper and more convenient, but it is still used for heating.

1.7 The Energy Crisis

At present most of the world's energy is supplied by coal, oil, natural gas and hydropower. Coal production could certainly be increased, but at great cost in damage to health and the environment. Hydropower is very useful, but is limited to around 10 per cent of world energy needs.

This leaves us with oil and its associated natural gas. In the last few decades many huge oilfields have been discovered and, in most cases, the oil is easily pumped out of the ground. It is clean and abundant and, until recently, it was very cheap as well. This is why oil has largely replaced coal as a source of energy in many countries. We have all become very dependent on oil, and this was brought home to us by the sharp rise in price in 1973.

Not only is oil increasingly expensive, it is fast running out. Oil wells last about twenty or thirty years, and so we can estimate future oil production from the present rate of discovery of new oilfields. The chilling fact is that this rate of discovery is steadily falling as more and more of the earth is explored. No new supergiant fields like those in Alaska, which supply the bulk of the oil, have been found since 1975. This means that oil production cannot go on increasing for very long. In a few decades it is bound to level off and begin to fall. The best estimates indicate that world oil production will peak early in the next century and then fall rapidly. Unless something very drastic is done there will be a serious energy shortage. Since it takes a long time to develop new sources of energy we must tackle the crisis now.

It has been estimated that to maintain present world oil production it is necessary to find new oil fields equivalent to the North Sea every two years. Geologists familiar with the poor rate of discovery during the last decade believe this to be impossible.

The seriousness of the impending world energy shortage is shown by estimates made in 1986 of the proven reserves of coal, oil and gas for the United Kingdom and for the world. When these are divided by the present rates of production they give the number of years supply listed in Table 1.

FUEL	UNITED KINGDOM	WORLD
Coal	114	125
Oil	14	34
Gas	16	58

TABLE 1. *Numbers of years' supply of fuel, estimated in 1986*

These figures refer to proven reserves, and so will be extended if further reserves are found. On the other hand, they are computed using present rates of production, and so will fall if production rises due to increasing demand.

The basic fact underlying the energy crisis is that world energy demands are rising rapidly and yet the production of oil, our main source of energy, will soon start to fall. There will be a widening gap between energy supply and demand. In the next few years the world will face an increasing shortage of energy, and this will have the most serious consequences if a new source of energy is not found. Not only will the price of oil rise, but the scramble for the remaining oil supplies will be a potent source of international tension. Already the Middle East is a highly unstable area, mainly because it is known to hold about half of the world's remaining oil reserves. The impact of the energy shortage is very different from country to country. Some are well-endowed with natural resources, while others are already seriously short of the energy they need.

The gap between world energy needs and resources is so large that it certainly cannot be filled by any one source alone. It is thus not a question whether this or that source will solve the problem. We need energy from all practicable sources, and we must do all we can to develop new sources. The questions to be faced are: which sources are likely to be economic; which can provide the large amounts we need; and which can do no more than provide useful supplementary amounts for particular purposes?

Even if we were able to satisfy all our energy needs by greatly increasing world energy production there would still be serious problems, in particular with the effects on the environment. It is obvious that world population and energy demand cannot go on increasing for ever. While we are rightly seeking to increase world energy production, we much also see how the demand itself can be moderated. There is no doubt that we waste a lot of energy, especially in the developed countries. We keep our homes too warm in winter, and use air conditioning too readily in summer. Many factories are using old and inefficient machinery, which wastes energy.

1.8 ENERGY CONSERVATION

At present, we all use energy very wastefully. Most of the heat in our houses escapes through the walls and the windows, and industrial processes, developed in times of cheap oil, are far less efficient than they should be. Huge amounts of energy are spent on pleasure in some parts of the world while, elsewhere, people lack the energy for the basic necessities of life

Considerable energy savings can be achieved at once simply by using less. By installing thermostats, lagging pipes and using double glazing we can maintain the same temperature with less fuel. Such savings are, however, relatively small and can come only gradually because the insulating material itself costs energy to manufacture, so it is some time before we recover the energy spent. While it is expensive to improve the thermal properties of old buildings, new ones can be designed to be energy-saving at much less cost. We may therefore expect buildings of all types to become gradually more energy-saving over the years, as old ones are replaced by new.

The motivation behind energy conservation remains extremely important, because it can significantly reduce energy demand. The present rate of increase of energy consumption in the richer countries is far too high, and cannot continue indefinitely. The style of life that has been unquestioningly accepted in times of cheap energy must be critically examined, so that the limited energy supplies of the world can be more justly shared.

A difficulty about improving energy efficiency is that it leads to a reduction in price. This is good if means that poor people are now able to afford the energy they need, but it may also mean that people who already have enough for their needs will spend even more on luxuries, and this could lead to increased energy consumption. This problem could perhaps be tackled by a system of differential tariffs, increasing the price with the level of consumption, so that the poor can afford what they need and the rich are discouraged from wasting it.

We must certainly conserve all the energy we can, and use what we need as efficiently as possible, but this is not enough to solve the immediate problems. Even with energy conservation the world energy demand will continue to rise, though not quite so sharply as before.

2

NEW ENERGY SOURCES

2.1 Renewable Energy Sources

Since we cannot get the energy we need from familiar sources, we must look for new ones. There are many possibilities, including solar, wind, wave, tidal and ocean thermal among the renewable sources, and nuclear power from the energy stored in the nucleus of the atom. How many of these are likely to be practicable?

The word renewable means that the energy source does not use up the earth's resources, and so hydropower is also a renewable source. Since, however, it is well-established, it is preferable to restrict the word to the new renewables listed above.

The sun pours enough energy on to the earth to satisfy our needs a thousand times. If we could find a practicable way of using it directly we would solve the energy crisis. Its heat stirs up the atmosphere and causes winds and waves, and these are sources of energy. We can also use the sun's energy directly, as solar power.

It is worth remarking that the sun is ultimately the source of all our energy, since coal and oil contain the sun's energy stored in the earth millions of years ago. Trees depend on the sun to live. The sun itself is a giant nuclear reactor and gets its energy from interlinked chains of reactions that convert hydrogen into helium, releasing energy. Eventually, the sun will burn out, and so energy from the sun is not indefinitely renewable. This need not worry us, however, since it will continue to shine for many millions of years.

Windmills have been used to grind corn since ancient times, and they are still widely used to pump water up from wells on farms. In such applications it does not matter much that the wind is not always blowing. It is easy to connect them to an electric generator to produce electricity. The new windmills are not like the old ones; they are large propeller blades mounted on high towers. About a thousand of these, spaced half a mile apart, are needed to equal the output of a coal power station. They are not reliable as a source since the wind does not always blow, and the cost of wind power is about three or four times that of coal power. Large arrays of windmills, called wind farms, are being built in the United States and in some other countries; this is only possible because they are heavily subsidised.

It is much the same for solar power. It is perfectly feasible to put a solar panel on the roof and let the sun's rays heat directly the water running through it. Solar ovens are useful for cooking meals in sunny countries. It is also possible to build huge mirrors to focus the sun's rays on to a boiler and to use the steam to generate electricity. But since the sun does not always shine this is unreliable, and it is far too expensive to be a practicable source of large amounts of energy.

On the average, the sun gives us about 200 watts per square metre at the earth's surface and conversion losses reduce this to about 50 watts per square metre; about enough to light one ordinary light bulb. There is nothing we can do to increase this.

We can easily work out the area of a collector required to satisfy the energy needs of an average family. Taking into account the efficiency of the equipment used, this has been estimated by Professor Hoyle to be about 1200 square metres. Thus a solar collector about the size of the Jodrell Bank radio-telescope would be required for each group of four families.

The tides are a practicable source of energy in a few places such as the La Ranche estuary in France and the Severn estuary in England. The water is held back by a dam at high tide and then allowed to run through turbines when the water level falls. A small tidal power station has been operating for many years at La Ranche and a detailed study has been made of the feasibility of using the Severn estuary. It is estimated that the energy it would produce would cost about twice that of a conventional power station. An investment of about fifteen billion pounds spread over about ten years would be needed before any power is produced, and the environmental effects are likely to be severe. There is also the difficulty that the power is not always produced at the time of day when it is needed, since the tides follow the lunar cycle.

The practicability of using wave power is much less certain, and various devices to harness the power of the waves are still in the experimental stage. There are formidable difficulties to be overcome, in particular the need to ensure that the device is strong enough to withstand storms and is not affected by corrosion. There is also the hazard to shipping and the problem of transmitting any power generated to the land. Wave power also severely affects the environment. Thus, at present, it seems unlikely that wave power will ever be economically practicable as a source of energy.

Ocean thermal energy is in a similar stage of development. It is thermodynamically possible to extract heat by using the difference in temperature between the surface and the depths of the ocean, but as yet there is no device that will do this reliably and economically.

Finally, geothermal energy is in much the same situation as tidal and hydroelectric; it is a well-established method that is making a useful contribution to world energy needs, but the number of sites where it can be developed economically is very limited. It cannot, therefore, ever be a major source of world energy.

All these renewable energy sources are either costly or unreliable, or available only in particular places on a small scale, so none of them can supply energy on the scale we need.

The main disadvantage of wind and solar power is that they are inevitably unreliable. Sometimes the wind blows and sometimes it does not, and clouds often obscure the sun. This would not matter quite so much if we could store the energy, but unfortunately there is no practicable way of doing this economically on a large scale. We have to generate the power when we need it, even on quiet, cloudy winter days. So if we went in for solar or wind power on a large scale we would still need to have coal, oil or something else as a standby. And since wind and solar are much more expensive than coal or oil, why use them at all?

The basic reason for the high cost of the renewable energies is very simple; the sun's energy is spread very thinly over the earth's surface, so we have to go to much trouble to concentrate it. It is very hard to break even when it comes to cost. It is only when nature does the concentrating for us, as it does with the fossil fuels and with hydroelectric power, that it becomes a practicable source of energy on a large scale.

Wind power is the most promising of the renewables, and many experimental wind generators have been built. It is important to continue this research to extend the range of economic applications.

All the renewable sources can only supply a few per cent of our energy needs. They are useful in a few applications, but are inherently unable to supply the bulk of our energy needs.

2.2 Nuclear Power

There is another source of energy that we must consider, the nucleus of the atom. In 1939, it was discovered that the nuclei of some very heavy atoms such as uranium 235 can break up into two nearly equal pieces, as shown in the Figure opposite. This process is called fission and the pieces are called fission fragments. They fly apart at high speed and this produces heat. Each fission also releases two or three neutrons, and these can trigger more fissions. This process continues very rapidly, with great release of energy. If the uranium is very concentrated the reaction takes place so quickly that there is a violent explosion. This is what happens in the atomic bomb.

It is also possible to dilute the uranium so that the energy is released gradually in a controllable way. This was first achieved by Fermi in 1942 when he built the first nuclear reactor. The energy of the fission fragments produces heat, and this can be used to boil water and drive a turbine to generate electricity, just as in coal and oil power stations.

Fermi's reactor was essentially a large pile of graphite blocks with rods of uranium in channels passing through the graphite.

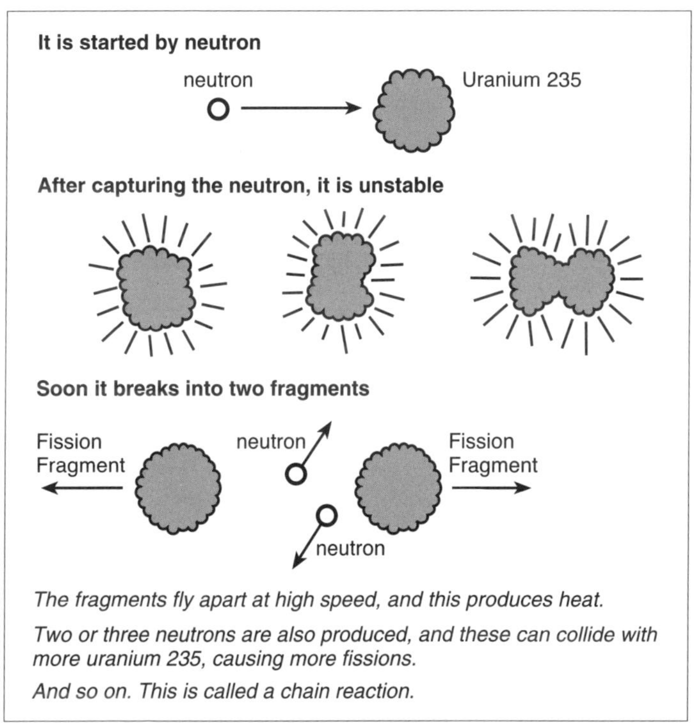

FIGURE 2. *Fission*

The graphite serves to slow down the neutrons so that they can cause further uranium nuclei to fission. About 2.5 neutrons are emitted on average from each fission, and many of these are absorbed by other nuclei or escape from the reactor. The chain reaction can only take place if, on average, at least one neutron from every fission causes another fission. If this number is below one the reaction stops, while if it is above one the number of fissions increases rapidly; we say that the reactor has gone critical.

The time taken from one fission to the next is so short that if these were the only disintegrations the reactor would be very difficult to control. Fortunately, however, some of the fission fragments also emit neutrons some seconds later; these are the delayed neutrons. The level of the reaction is then adjusted so that it only becomes critical with these delayed neutrons, and then it is easier to control. The intensity of the neutrons inside the reactor, and thus the power level, is controlled by rods containing a material that absorbs neutrons. These rods can move in and out of the reactor. When they are inserted, the reaction cannot take place. To start the reactor, they are moved out slowly until the reactor becomes critical with the delayed neutrons. The power then rises to the desired level, and then the rods are pushed in again until the multiplication factor is reduced to unity, so that the power level remains steady.

As the reaction proceeds, fission fragments are formed and remain in the uranium rods. As they can absorb neutrons, they slow the reactions down, so that to maintain the power level the control rods are moved slowly out. Eventually there are too many fission fragments in the reactor, so the rods are removed and replaced by new ones. This can be done continually as the reactor is operating. The fission fragments are highly radioactive and so they must be prevented from escaping and harming the environment. The spent rods containing the fission fragments are treated chemically to separate the uranium, which can be used again, and the plutonium, which is produced during the operation of the reactor. Plutonium is also fissile and can be used in a reactor. The separated fission fragments constitute nuclear waste, and they are stored as described in Chapter 5.2. The place where the separation takes place is called a reprocessing plant, and one of these is THORP at Sellafield, in Cumbria.

A nuclear reactor is a new source of energy, and it uses a material, uranium, for which there is no other use, except on a very small scale to colour glass. It must therefore be assessed critically to see if it can supply our energy needs.

There are now over four hundred nuclear power reactors in operation in many countries worldwide, and together they supply about 20 per cent of the world's electricity. The number in each country depends on the availability of other energy sources, especially of coal and oil. In France, for example, a highly industrialised country with practically no indigenous coal or oil, nuclear power is now responsible for over 75 per cent of the electricity generated. In Western Europe, nuclear power has overtaken coal, and supplies about 50 per cent of the electricity. In many other countries a very substantial amount of the electricity is generated by nuclear reactors, as shown in Figure 3. There is thus no doubt that nuclear power has the capacity to supply a large fraction of world energy needs.

The relative costs of coal and nuclear power depend on a number of factors, such as the proximity of coalfields, and detailed studies have shown that in some cases they are about the same and in others coal is almost twice as costly as nuclear. The basic reason for the relative cost advantage of nuclear power is that the energy is extremely concentrated. There is as much energy in a pound of uranium as in a thousand tons of coal. To some extent this advantage is reduced because nuclear power stations are more costly to build, but the running costs are less because such small amounts of fuel have to be transported to the power station.

The remaining question is whether there is enough uranium at an economic price to fuel the nuclear reactors that will be needed in the next century. Uranium is very widespread, but the concentration of uranium varies from one deposit to another. The amount we can extract depends on the price we are prepared to pay for it, since the higher the price the more ores are economically workable. There is certainly enough uranium to last for several decades at the present rate of production, and as this is exhausted thorium can also be used.

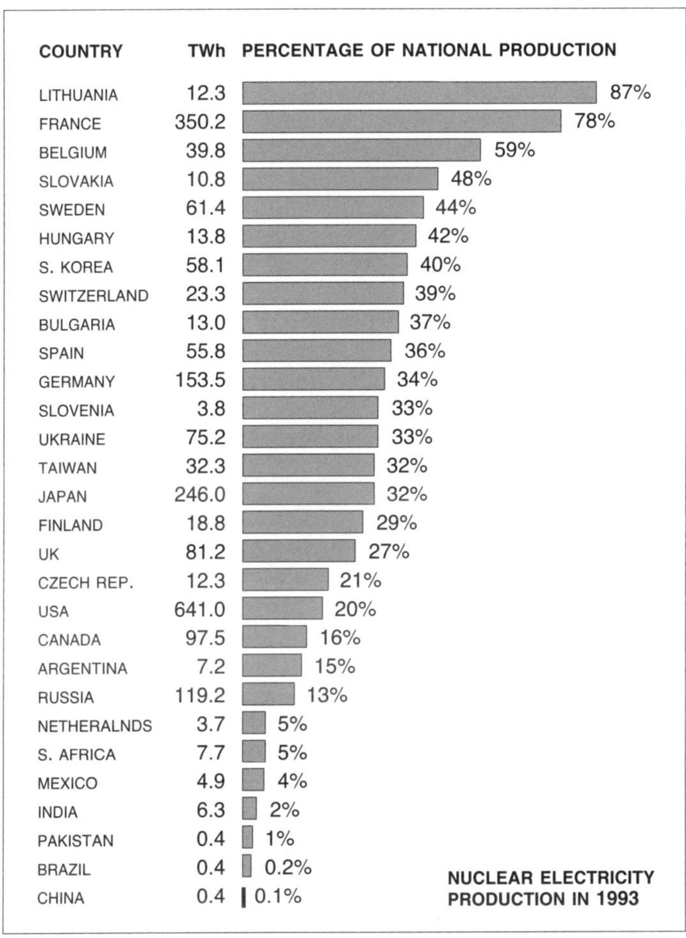

FIGURE 3. *Percentage of electricity generated from nuclear power in 1993 in several countries. The second column gives the total annual production.* Nuclear Issues *Vol. 16/3 March 1994*

It that were the whole story, the development of nuclear power would certainly ease the energy crisis, but would not solve it. In about forty or fifty years, when all the useable uranium and thorium is exhausted, we would once again face a severe shortage of energy.

Fortunately there is another way of burning uranium, known as breeding. Natural uranium consists of two types of uranium nuclei, and only one of these, uranium 235, is easily fissile. This is present in natural uranium in a very small proportion, less than one per cent. The present nuclear reactors, called thermal reactors, burn only this amount of the uranium. It has now been found possible to design a reactor, called a fast reactor, that is able to burn the remaining 99 per cent of uranium as well. Allowing for losses, this effectively multiplies the amount of fissile material available by a factor of about fifty. Several large fast reactors have already been built, and they are ready to take over power production when the price of uranium makes it economical to do so.

There is thus enough fissile material available for the forseeable future. It is no use thinking more than about fifty years ahead, because before then there may be new technological developments that completely transform the situation. It is also quite possible that by then the problem of using the fusion reactor to generate power economically will have been solved.

It has long been known that if the heavy isotopes of hydrogen with masses two and three, called deuterium and tritium, are brought close enough together they react and release a huge amount of energy. For this to happen, they must be raised to a very high temperature so that the increased thermal motion overcomes their electrical repulsion. In the hydrogen bomb this is done by using the fission reaction as a detonator. For peaceful purposes this must take place in a controlled way.

Several experimental devices have been built in the United Kingdom, the United States and Japan to study how the fusion reaction can be controlled. The problem is much more difficult than was at first thought. The method used is to heat electrically the mixture of deuterium and tritium until it becomes a very hot gas called a plasma. The difficulty is to keep the plasma still long enough for the fusion reaction to take place. Scientists are using strong magnetic fields to confine the plasma, and recent experiments have succeeded in producing significant amounts of power. It is confidently hoped that the next generation of experimental machines will reach the break-even point, when more power is produced than is used to heat the plasma.

Anticipating this, possible designs for a full-scale fusion reactor are now being studied, so that they can be built when the reaction is sufficiently understood and it is economical to do so. It is unlikely that this will happen earlier than about forty or fifty years from now.

Fusion power is so important because deuterium is present in seawater, so it can provide practically unlimited energy. The tritium that is also needed can be produced by the fusion reaction, and so can be continually replenished.

2.3 Safety

We must also consider the question of safety. There is no completely safe way to produce energy. Coal mining is notoriously dangerous, oil wells catch fire, tankers collide or explode and dams burst. The renewable energy sources are sometimes described as safe or benign, but if we take into account the risks involved in making the materials, and then constructing them, it turns out that they are not so safe after all.

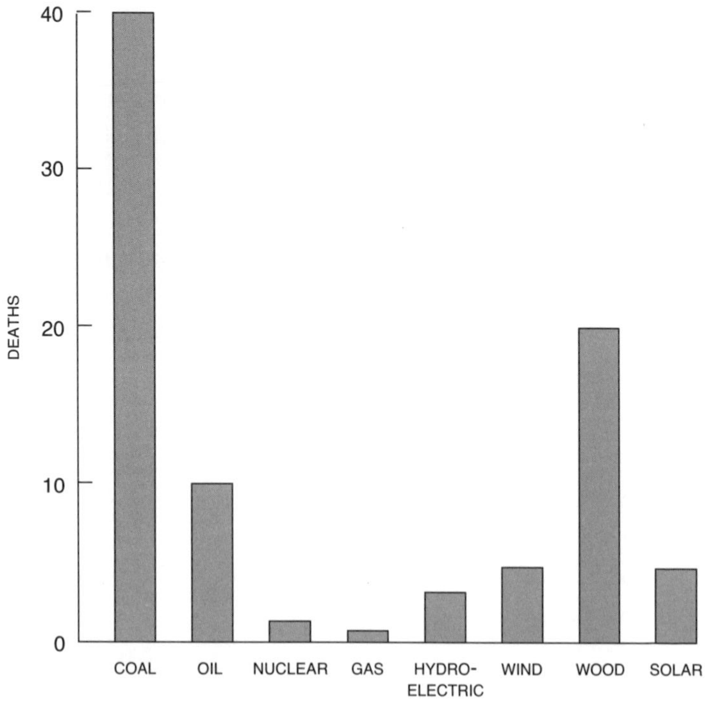

FIGURE 4. *Average numbers of deaths associated with the production of 1000 megawatt years of electricity in different ways (a megawatt is a million watts) (Inhaber1981)*

To assess safety objectively we must take into account all the risks for each energy source. Mining, transport, construction, operation, maintenance and distribution all involve risks. Some are direct and affect the workers, and some are indirect, like pollution, and affect the whole population. To compare the sources of energy these risks must be added together. The results of a recent study are shown in Figure 4. This gives the risks associated with different energy sources in the form of deaths involved in the production of a thousand megawatt years of electricity in different ways. Rather similar results are obtained for the injuries to the workers. There is much argument over these estimates, and the results should not be regarded as final. However further studies are likely to give the same ranking order for the various energy sources.

At first sight some of these results might seem surprising. Coal is very hazardous because of the dangers of mining and also because of the large amounts of sulphurous and nitrous gases discharged into the atmosphere. More than 1500 seamen lost their lives in tanker accidents from 1968–79. Wind and solar are more dangerous than we might expect because very many collectors have to be built, and this means mining and construction hazards. Nuclear comes out quite well because uranium is such a concentrated power source that relatively small quantities need to be mined, and also because there is no poisonous smoke from nuclear power stations.

It is notorious that it is the spectacular accidents that make the headlines and capture the public imagination, although over a long period of time they contribute relatively little to the total hazard of a power source. This is very familiar in other contexts; a train crash causing some tens of deaths receives more attention than thousands killed individually in road accidents.

FUEL	NUMBER	TYPE	DEATHS PER ACCIDENT	AVERAGE PER YEAR
Coal	62	Mining	10-434	>200
Oil	63	Fire	5-500	~50
Gas	24	Fire	6-452	>80
Hydro	>8	Dam burst	11-2500	>200
Nuclear	1	Chernobyl	31	-

Table 2 Severe Accidents 1969–1986

Such spectacular accidents are also a tragic feature of energy generation and some of the most notable in the period 1969–1986 are listed in Table 2. Mining accidents are particularly serious, and over 80,000 miners were killed in accidents from 1873 to 1938, and many hundreds of thousands more had their health permanently impaired by silicosis and other diseases.

The hazards of energy production can, of course, be reduced by increasing the safety precautions, but this inevitably increases the cost. For example, many of the poisonous gases can be removed from the effluent gases from coal power stations by installing special equipment, but this is very expensive. In the end we have to strike a balance, and this can only be done by carefully evaluating the hazards of all the energy sources.

We know a great deal about radioactivity because extremely small amounts can be detected by special instruments. In this way we can find out how it travels through the atmosphere, the rivers and the seas, and our own bodies. Very detailed and extensive studies of nuclear hazards have been made, and this allows strict limits to radiation doses to be established and enforced. This will be considered in more detail in the next Chapter.

Much less is known about many other hazards of life, particularly those associated with the harmful chemicals released into the atmosphere by coal power stations and many factories. These may travel thousands of miles and fall as acid rain. Other chemicals are released into rivers and seas. We are all familiar with the black smoke that hangs over industrial cities and with polluted rivers and lakes. All this exacts a heavy toll in disease and in shortening of life, as well as killing trees, plants and fishes.

The main components of acid rain are sulphur dioxide, nitrogen oxides and hydrocarbons. Sulphur dioxide damages limestone and marble and some types of sandstone, and so is harmful to buildings. Together with nitrous oxide it can be lethal to fish. Already this is affecting lakes, rivers and forests in many countries, including Scandinavia, Austria, France, Switzerland and the United Kingdom. The acid in the water can dissolve copper and lead from pipes, and cadmium and mercury from the soil, and all this is hazardous to health. More sulphate particles in the air are increasing the levels of asthma, allergies and hay fever.

These noxious gases are emitted in many industrial processes, but power plants contribute a large share. Analyses of emissions in the United Kingdom showed that 66 per cent of the sulphur dioxide came from fossil fuel power stations. About 46 per cent of the nitrogen oxides came from the same source.

To sum up the comparison between the different energy sources so far, we have seen that only coal, oil and nuclear have the capacity to provide the huge amounts of energy needed by large cities and modern industrial societies. Hydropower is important but limited by the availability of suitable rivers. The renewables like wind and solar have useful small-scale applications, but cannot provide the reliable and economical power that we need.

3

ENVIRONMENT AND HEALTH

3.1 Energy and Health

Our health and well-being and our standard of living depend directly on our environment. By our environment we mean our surroundings, the air we breathe, its purity and its temperature, and the noises and other disturbances to which we are subjected. We mean the food we eat, the comfort of our homes and the availability of convenient transport and communication. We include also the appearance of the city or town where we live, and the beauty of the surrounding countryside. All this determines our bodily and mental health.

Much of this environment depends on energy, and so energy, by contributing to our environment, also affects our health. The relation between energy and health is shown in Figure 5. The vertical scale is the average life expectancy, a crude but effective measure of health, and the horizontal scale shows the average annual consumption of energy by each person. Each dot shows the relation between life expectancy and energy consumption for a particular country. In the lower left-hand corner are the poor countries of the world, with little energy and low life expectancy. At the top right hand are the rich countries of Europe, North America and Japan, where each person uses up tens or hundreds of times the energy per person of the poorer countries, and where the life expectancy is about twice as great.

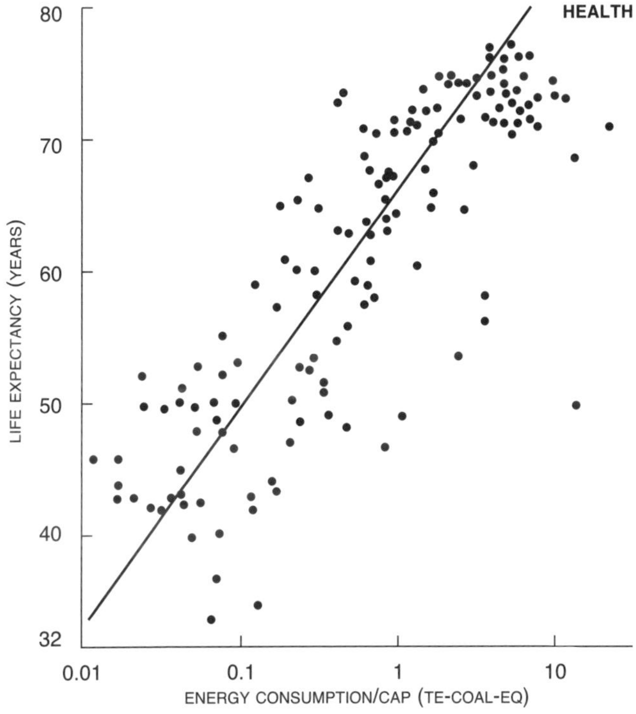

FIGURE 5. *Life expectancy 1980-85 vs energy consumption/person/year, 150 countries.*

This shows that if we want to increase the standard of living of the poorer countries of the world we must increase the amount of energy available to them. Unfortunately, however, the energy sources themselves, the power stations and associated industries, all have a deleterious effect on the environment. It is therefore important to examine the effects on the environment of the various energy sources in order to decide which of them are preferable from this point of view, and how their environmental impact can be minimised.

Some of these effects are damaging to our health and others, to be considered in the next chapter, are destructive of the landscape. The waste from power stations and the ways to deal with it are described in Chapter 5.

3.2 Pollution by Coal and Oil Power Stations

A major effect of power stations on the environment is the pollution of the atmosphere and the earth. All power stations pollute in one way or another, and the more power we generate, the greater the pollution. In some cases the pollution is an inevitable result of the process used and therefore cannot be prevented, while in others it can be controlled so that there is no harm to people.

Coal power stations are particularly polluting, and a typical one will emit each year eleven million tons of carbon dioxide, sixteen thousand tons of sulphur dioxide, twenty-nine thousand tons of nitrous oxide, a thousand tons of dust and smaller amounts of a whole range of other chemicals, including aluminium, calcium, potassium, titanium and arsenic. All this is poured into the air we breathe and it damages our health. It travels large distances and falls as acid rain. This has poisoned lakes in Scandinavia so that the fishes die, and is already destroying the forests of central Europe. It is possible in principle to remove some of these poisons before they are expelled into the atmosphere, but to remove even a large fraction of them would be prohibitively expensive.

Oil power stations are not quite as polluting as coal power stations, but they inevitably produce the same amount of carbon dioxide. Oil does have an additional hazard due to the need to transport it over large distances. Overland this can be done by pipelines, and these affect the environment not only visually but due to the possibility of leaks. The seasonal migration of animals such as caribou may also be affected. Much oil is also transported across the oceans by huge supertankers. This is very efficient, but if they run on to the rocks and are holed or break up large quantities of oil are released. This oil floats on the water and can travel large distances, causing a major environmental disaster. The *Torrey Canyon*, *Exxon Valdez* and, more recently, the *Sea Empress* are still vivid memories. The oil fouls the beaches and kills birds, marine life and fishes.

The oil can be dispersed by spraying with chemicals, but in some cases this does even more environmental damage. Although some of changes may be irreversible, the environment often recovers remarkably quickly, so that some years later few traces of the disaster remain. As in other matters, it is important not to overstate the extent of the damage, as is often done by environmental organisations and by those who justifiably seek compensation. Oil tankers are subject to stringent regulations, and improvements in their design are continually being made. Nevertheless these disasters still happen and remain a serious environmental hazard of oil power.

3.3 Pollution by the Renewables

The renewable energy sources are widely believed to produce no atmospheric pollution. This is true for their everyday operation, but not for their construction. Like all power stations, they have to be made in factories, and this inevitably produces pollution. This is by no means small, as very large numbers of windmills and solar panels have to be made to equal the output of a conventional power station.

3.4 POLLUTION BY NUCLEAR POWER STATIONS

Since nuclear power is the only large-scale source, together with coal, that can take over from oil and natural gas, its effects on the environment must be carefully examined. In normal operation, a nuclear power station releases only very small amounts of radioactivity into the atmosphere. These amounts are even less than the radioactive emissions from coal power stations.

To put into perspective the radiations from the radioactive material produced in nuclear reactors they must be compared with other sources of radiation. We are constantly exposed to the same types of nuclear radiations from radioactivity in the earth and from the cosmic radiation that enters the earth's atmosphere from outer space. This natural radiation amounts to about 1 millisievert (mS) per year, of which about 0.3 comes from the cosmic radiation, 0.4 from the soil and the air, and 0.3 from the radioactive material in our own bodies. (A sievert is a unit of radioactive exposure). In addition, we receive on the average about 0.45 mS from X-rays and other forms of medical treatment and about 0.02 mS from the fall-out from atomic bomb tests. The radiation from nuclear power reactors is kept carefully under control and is currently about a quarter of a millirem per year. The relative proportions of these contributions to the radiation we receive are shown in Figure 6.

RADIATION EXPOSURE OF THE UK POPULATION

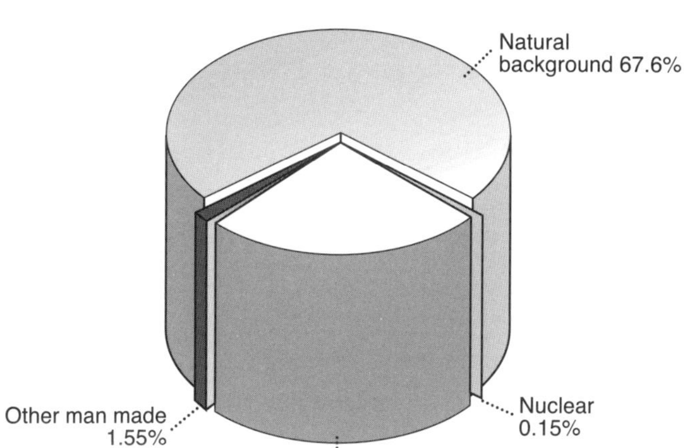

FIGURE 6. *Percentage contributions to the total radiation exposure of the population of Britain (*Atom, *January 1980 p.4)*

All of these radiation levels are far below those that could cause any sort of illness. We know this because in some parts of India and Brazil the natural background is over ten times the average quoted above, due to the presence of radioactive rocks, and even in these areas the population shows no signs of additional radiation damage that could be attributed to this extra natural radiation. Professor Fremlin from Birmingham University has recently estimated that the radiation dose received in Britain due to the nuclear power stations reduces our average life expectancy by one or two seconds. The radiation dose is similar to that from coal power stations, since all coal is contaminated by uranium.

There have been widespread fears about the genetic effects of nuclear radiations. This has been studied by looking at the children of the survivors of Hiroshima and Nagasaki, who all received massive doses of radiation. These children show no differences from normal children, so fortunately there is no evidence for such genetic damage.

Nuclear power stations produce radioactivity in the form of fission fragments. This constitutes nuclear waste, and the methods used to get rid of this waste safely are described in Chapter 5.

3.4 REACTOR ACCIDENTS

There is also concern at the possibility of reactor accidents that could release radioactivity directly into the atmosphere. However carefully a reactor is designed there is always the fear that something quite unexpected might happen, or that the operators might make a serious error, leading to a disastrous explosion. This fear is given substance by the many accidents that have taken place, in particular those at Brown's Ferry in 1975, at Three Mile Island in 1979 and, worst of all, at Chernobyl in 1987.

A reactor accident can occur when the power level rises far above the designed level. The reactor then expands, increasing the loss of neutrons through the surface and thus slowing the reaction. Reactors that behave in this way are thermally stable, because a rise in temperature tends to reduce the power level, thus automatically lowering the temperature. A dangerous rise in temperature also triggers the control rods, containing neutron absorbing materials, which can be shot into the reactors very rapidly to stop the chain reaction.

Even if the nuclear reaction is stopped in this way, the heat continually given out by the fission fragments in the fuel rods could melt the uranium and then the steel containing vessel. If this fractures, large amounts of radioactivity would be released. If the reactor building is also damaged, the radioactivity could be released into the atmosphere, resulting in a serious health hazard.

Normally, such a rise in the temperature of the reactor core is prevented by the circulation of cooling liquid, but this may be interrupted if the pipes are broken. To cope with such an emergency there are usually duplicate cooling systems and provision to inject more cooling fluid whenever the pressure in the primary system falls below a safe level.

The accidents at Brown's Ferry and at Three Mile Island illustrate the results of faulty design and faulty operating procedures. At Brown's Ferry the trouble started when an electrician was testing for air leaks in the cable room beneath the main control room. He used the standard test for leaks, a lighted candle, whose flame is sensitive to air pressure. There was indeed a hole, and because the air in the reactor containment building is kept at a lower pressure than outside (to prevent the escape of radioactivity) the flame was sucked into the hole. There it set fire to strips of polyurethane foam inside the hole. This in turn set fire to the insulation of the electric cables, many of them controlling the reactors safety system. This made it very difficult for the reactor engineers to cool the reactor down. Eventually, they succeeded in doing this sixteen hours after the fire had started by using the auxiliary water supply. The fire itself was out of control for over seven hours and damaged sixteen hundred control cables causing millions of dollars of damage. After that experience flammable material is never used for cable insulation and there are two independent sets of cables to supply power to the safety equipment. Fortunately there was no release of radioactivity in this accident.

The accident at Three Mile Island was of a different character. The first thing to happen was a breakdown of the pumps circulating water in the secondary cooling system. The water in this system circulates near the water from the primary system that comes from the reactor and removes some of its heat. Since it was no longer being cooled, the water in the primary system heated up, together with the reactor itself. Emergency cooling systems should then have come into action, but the valves were closed, which was not known at the time. The pressure in the primary water system was then reduced automatically by a valve above the reactor, which failed to close. Additional water was then pumped in automatically to replaced the primary water. The operators then misinterpreted a dial reading and, thinking that it indicated that there was no vapour phase above the cooling water in the pressuriser, stopped the emergency water supply for fear of filling the system with liquid, which, on expansion, would have burst the containing vessel.

The top of the reactor core then became uncovered and its temperature rose rapidly. The fuel rods buckled and cracked, and the zirconium casing reacted with the water to give free hydrogen, which collected in a bubble at the top of the rector vessel. There was then some concern that the hydrogen would react explosively with any oxygen that might be present from the water molecules, bursting the reactor vessel and perhaps also fracturing the containment building, allowing a massive release of radioactivity. Actually, there was never any danger of this happening because the zirconium combines with any oxygen that night be present, but this as not realised at the time and several alarming announcements were made to the public before the reactor was eventually brought under control.

Only a small amount of radioactivity was released, enough to give people living within fifty miles the dose (about 0.01 mS) that they would receive in one day from natural sources. The major health effect was the distress caused by the spread of alarming and inaccurate information.

There were some widely publicised stories about a 50–100 per cent rise in infant mortality in the Harrisburg area after the accident, and predictions of 4000 to 8000 extra cancer deaths in the following decades. In fact, the actual infant mortality statistics showed no change associated with the accident, and the radioactivity released is unlikely to cause even one cancer death.

The accident at Chernobyl was, without any doubt, a major disaster. Indeed, in a recent Oxford lecture Zhores Medvedev, the Russian dissident and onetime Head of the Institute of Radiological Biology in Obninsk, said that it was one of the main factors leading to the collapse of the Soviet Union. A short time before Chernobyl, Mr Gorbachev's energy policy for the Soviet Union put increasing reliance on nuclear power. After Chernobyl, local anti-nuclear pressure stopped nearly all further construction and thus gravely exacerbated the economic difficulties that already faced him.

The Chernobyl accident was caused by a combination of design and operator failures. It was thermally unstable at low power, which, by itself, is not dangerous since it can be corrected by controls and by never running at low power. Another unsatisfactory feature, however, is that the neutrons are slowed down by the cooling water as well as by the graphite. The water also absorbs neutrons and so if the amount of water is reduced, for example by vaporisation, the neutron absorption is also reduced and this destabilises the reactor. Neither of these features would be permitted in a Western reactor. The accident happened during an experiment intended to find out whether the electricity generator would continue to supply electricity for a short time when it was no longer supplied with steam. It was thus an electrotechnical experiment, and the minds of the operators were on the electricity generating equipment and not on the reactor. The experiment was directed by an electrical engineer, not by a nuclear scientist.

They did not want the reactor to cut out during their experiment, so they disconnected the emergency core cooling circuit. The power fell to near the danger level, so to maintain power, the control rods were pulled out, putting the reactor into a dangerous condition. Additional cooling pumps were switched on, increasing the amount of water in the reactor and hence absorption of neutrons. To prevent the reactor cutting out, the operator switched off the emergency shutdown mechanism and prepared to begin the experiment. Then the power started to rise rapidly and he ordered an immediate shutdown. The control rods could not move fast enough to stop the power rise. The reactor went critical and in a few seconds the power level increased to a hundred times full power. The fuel rods became extremely hot and burst open, vaporising the surrounding water. The steam pressure blew off the top of the reactor and sent a shower of blazing graphite into the air. The graphite left in the core caught fire and burned fiercely for several days, sending a plume of radioactive smoke high into the atmosphere.

Firefighting teams raced to the reactor and with great courage tried to contain the damage. Thousands of tons of boron, clay and lead were dropped by helicopter into the blazing reactor and eventually the fire was put out. The workers in the immediate vicinity of the reactor at the time of the accident received large radiation doses, and 31 died soon after. In addition, 145 people received large but not lethal doses; they are being carefully monitored and it is likely that some of them will develop cancers over the coming years.

Some of the radioactivity released was deposited on the ground in the region surrounding the reactor, and some was blown large distances and was detected in many European countries. To minimise the danger to their health, the authorities evacuated over a hundred thousand people from the region extending about thirty kilometres around the reactor.

To assess the damage caused by this radioactivity two questions must be answered. Firstly, how seriously is the surrounding region contaminated, and will it ever be safe for the inhabitants to return? Secondly, what will be the effects of the much smaller doses received throughout Europe and elsewhere?

It is easy to detect the radioactive decay of a single nucleus, so exceedingly minute amounts of radioactivity can be measured. It is therefore already well known that practically everything is radioactive to some extent, including our own bodies. We are continuously irradiated by the cosmic radiation from outside the earth, and by the radioactivity in the ground. This constitutes the natural background of radiation, and provides a standard with which to assess the likely effects of any additional radiation due to the nuclear industry, to medical irradiation and to accidents like Chernobyl. Furthermore, the natural background varies very markedly from place to place, depending on the geology of the region. In Cornwall, for example, the natural background is over twice the average for Britain, due to the granitic rocks that contain uranium.

Measurements of the radiation levels around Chernobyl show an excess over the natural level attributable to the accident. Detailed maps have been made, and show a very irregular distribution, impossible to summarise concisely. All that can be done is to gives a few numbers to indicate the general levels found. The following numbers are for levels of radiation expressed in terms of the unit microsieverts per hour.

The highest readings were, of course, near the reactor itself, being 30–50 at 300 metres. Further away, in the town of Pripyat, they were 0.5 to 0.9, and at the edge of the 30 km zone around the reactor, 0.7. In an administration building 18 km from the reactor, 0.2. For comparison, the Swedish scientists who made these measurements in September 1990 found 1.9 on the plane to Moscow. The average British exposure from natural radiation is 0.25, and that in Cornwall 0.85. From such figures it can be concluded that apart from the immediate vicinity of the reactor, the radiation levels are similar to those in Cornwall, or less.

This picture is confirmed by observations of the flora and fauna. Doreen Stoneham from the Oxford Research Laboratory for Archaeology visited the area in 1990 and found deserted gardens flourishing, with fruit trees laden with apricots, cherries, apples and greengages, and an abundance of flowers, including rosebay willowherb, wild chicory, thyme, cow parsley and verbascum. Birds were nesting in the chimneys of abandoned houses and in one nest there were two young storks. The only obvious damage was to the older pine trees.

Pripyat was derelict, with creepers growing over walls and pavements. Inside buildings the doses averaged 0.25, but outside they were much higher: 3-8 on the grass, 12 on exposed brick and 110 on another grass verge. She concluded that the contamination of the bricks is a difficult problem, but the biggest obstacle to re-occupation is the deterioration of the buildings themselves after years of neglect.

The health of the million or so people still living in 2700 settlements in the contaminated area was studied by 200 international experts from 22 countries co-ordinated by the International Atomic Energy Agency. The results were published in 1991 in an 800-page technical report. The project was led by the Director of the Hiroshima Radiation Effects Foundation and included members of the World Health Organisation, the International Labour Organisation, the UN Scientific Committee on the Efects of Atomic Radiation and three other independent organisations. It is thus difficult to accept the claims of certain environmental groups that their report was a political whitewash and that they were duped by the Soviet authorities. The report found no health disorders directly attributable to radiation exposure and, in particular. no indications of an increased incidence of cancers, including leukaemia. This was not surprising, because radiation-induced cancers often take years to develop.

To establish the presence of any adverse health effect, it is necessary to compare the observed incidence with that expected in the absence of the accident, making allowance for any changes in lifestyles and also in the diagnostic techniques. Since many people were irradiated, it was expected that there might be a detectable increase of some cancers after about seven years. In particular, some population groups received thyroid doses around 1 Sv, and several hundred cases of childhood thyroid cancer were indeed found during 1991–94 in Belarus and the Ukraine, rather sooner than expected. The rates were so greatly in excess of those occurring before the accident that they cannot be ascribed to improved diagnostic techniques. There was no evidence of any other effects, particularly of childhood leukaemia, congenital abnormalities or any other radiation disease either in the contaminated regions or in Western Europe.

It has also been reported that there have been 6000 deaths among the workers engaged on the cleanup operation, but this is just about the number that would have been expected from other causes in the very large number of people involved (some 800,000) over the period in question. Many of these workers suffered from radiation sickness.

The psychological effects of the accident were widespread and severe. Nuclear radiations are unseen and the information given to the public was not enough to allay their fears that they had been subjected to dangerous radiation. They had been told that such an accident was impossible, and yet it happened. Thousands of people were uprooted from their homes and told to avoid certain familiar foods. The official reactions, understandably enough, were not always consistent. It was the time of glasnost and perestroika, and Chernobyl came to symbolise all that was wrong with the old system. In this situation there was a tendency to attribute every illness to the effects of Chernobyl, whether or not it is of a type that can be caused by radiation. To deal with this situation, every sufferer from leukaemia was officially classified as a 'victim of Chernobyl', and is entitled to compensation, although there is no evidence that the illness is actually due to Chernobyl. Groups of children who have been sent to recuperate in other countries have been found to suffer from malnutrition, but not from the effects of radiation.

This conclusion is supported by the results of Professor Kellerer from Germany, who found that the people in the Chernobyl area thought that their medical conditions were the result of radiation, and did not understand 'that the undoubted deterioration was the result of changed lifestyles and an enormous degree of anxiety, a sort of self-amplifying problem. People believe they are surrounded by problems and they do not drink local milk, they abandon their cattle, rear no poultry and keep children indoors. This whole thing leads to a breakdown of lifestyle. So quite understandably you have indeed increased morbidity.'

Undoubtedly there was much suffering due to the evacuation itself, and even more due to the very natural fears of the effects of radiation, not helped by the media. Indeed, the international project commented: 'The psychological problems were wholly disproportionate to the biological significance of the radioactive contamination. The consequences of the accident are inextricably linked with the many socio-economic and political developments that were occurring in the USSR. A large proportion of the population have serious concerns. The vast majority of adults examined in both contaminated and control settlements either believed or suspected they had illness due to radiation.'

This was underlined by a report from four Swedish scientists who visited the area in 1990. They were distressed to find evidence of widespread ill health, but noted that this was not induced by radiation. They believe that 'many people are exploiting the biggest consequences of the Chernobyl accident – the radiophobia – to further their own aims.' Before the disaster there was chronic malnutrition, and now people are afraid to eat or go outdoors.

Soviet experts are well aware of this, but their reassurances are undermined by local officials with their own political interests. The Swedish scientist interviewed Professor Guskowa, a member of the Soviet Academy of Science and noted: 'When we took our leave of Angelina Guskowa we said goodbye to a very pessimistic woman who all her professional life has been engaged in work at home and abroad to cure people suffering from the effects of radiation, to spread information about these effects, and who has endeavoured to estimate the risks of radiation. It was obvious that she was disappointed in how the egotistical interests of politicians and suchlike who, totally without support for such a claim, pose as experts, have managed to sabotage everything she has worked for.'

What of the effects of the radioactive contamination of Europe as a whole? Here again the deposition was extremely patchy, depending on the winds and rain. Saltzburg even received more than Kiev. The additional radiation received by people in Europe ranged from the insignificant to around the level of the natural background, so it is not easy to estimate the effects on health. The effects of large radiation doses are well known, but almost nothing is known for certain about the effects of very small doses. It could be that the body is able to repair any damage due to very small doses. There is even some evidence that very small doses are beneficial to health. In Cornwall, for example, where the natural radiation background is higher, the incidence of leukaemia is rather less than in the rest of Britain. Nevertheless, in spite of this, it is usual to assume that the effect of radiation is strictly proportional to the dose. This is almost certainly a very pessimistic assumption.

Making the assumption of proportionality, it is easy to obtain figures like the 40,000 deaths worldwide over the next 50 years quoted in the media. This would be statistically undetectable among the 500 million people expected to die of cancer in any case in the same period.

Once again the natural background provides a standard for judgement. The extra doses due to Chernobyl are very small compared with other increases we accept without hesitation, such as those on airplane flights, on mountains (due to the cosmic radiation, which becomes more intense with increasing height) and due to the variations from one place to another. There has been discussion about whether more people should be evacuated from the Chernobyl neighbourhood, but none about the evacuation of Cornwall, which has a higher natural radiation level.

There were even some reports of radiation effects beyond Europe. A much-publicised news item reported a strong correlation between mortality rates and the radiation levels from Chernobyl dust in the United States. This looked very convincing, until one reflected that if such minuscule amounts of dust could have such pronounced effects there then the much larger (though still very small) amounts received in Europe would have spectacular effects, which were not present. Examination of the basic data showed that the alleged correlation was spurious. By that time the media had lost interest in the story and had gone on to the next radiation scare.

After Chernobyl several countries, including Sweden and Switzerland, voted to phase out nuclear power as soon as practicable. But when they looked into the alternatives they realised that they were even less attractive. Coal is seriously polluting and is contributing to acid rain and to the greenhouse effect. Oil will become more expensive and it is politically undesirable to rely on it. Hydroelectric power is already used to the practicable limit in most European countries, and the renewables such as wind and solar are not credible as a large-scale source. So the resolutions to phase out nuclear power are being quietly forgotten.

The whole story of Chernobyl is one of tragic and misguided muddle that brought death to some people and suffering to many hundreds of thousands of others. It is important to examine it carefully, with due regard for the facts, to see what lessons can be drawn for the future.

The nuclear physicists and engineers have developed a new source of power that has many advantages over the alternatives as regards availability, capacity, safety, cost and effects on the environment. But like most modern technologies, it is complicated and has to be handled with due care. This is not confined to nuclear power; it applies to all industrial processes and modes of transport. The choice is ours, both individually and as members of society. We do not entrust a car, let alone an airliner, to one who is not properly trained; if we were to do so we would be asking for trouble. In the case of new technologies, it is the duty of governments to lay down and enforce strict rules of operation. Indeed, nowadays, when the effects of a major disaster are felt worldwide, it is arguably the duty of the United Nations to do this.

At this point the duty of the scientists and engineers ceases, and that of the politicians begins. It is now clear that the politicians in the Soviet Union failed in this duty. The reactors were built hurriedly for the dual purpose of producing weapons grade plutonium and providing power for civil use. To achieve this, a design was adopted that would never have been accepted in the West. As already mentioned, it was basically unsafe, being thermally unstable. To prevent a dangerous situation developing, strict operating instructions were laid down. Unfortunately people do not always obey instructions, and on the night of the disaster the reactor was run at low power so that the operators could carry out a certain experiment. The reactor was designed to shut down automatically if it was operated in the unstable region, but since this might spoil their experiment, the operators switched off the safety devices, and disaster followed. There was no one in the control room who understood the risks that they were taking. None of this could have happened if the design and construction of reactors had been carried out under the control of an international regulating body.

After the disaster, the Soviet officials imposed secrecy until the radioactive fallout in other countries forced them to admit that it had occurred. The evacuation of people was probably essential in the circumstances, but not enough was done to give them accurate knowledge concerning the hazards associated with the doses they received.

Local politicians reacted by using the disaster to advance their own political ambitions. Those in other countries reacted with panic and voted to phase out nuclear power without a proper study of the implications. The people most affected who lived in the area around the reactor were not given accurate information about radiation hazards, and so they were terrified by the inflated stories abounding in the press and many of them suffered severe psychological problems. They became apathetic and feared to eat readily available food. In the end the danger to health from this were far larger than the dangers attributable to radiation.

Soon after the accident, a massive clean-up operation was started and has continued with international assistance. Extensive measurements have been made of radiation levels to determine when it is safe for the population to return. A major problem remains the healing of the deep psychological wounds.

And what of the Chernobyl-type reactors, not only in what was the USSR but also in Eastern Europe? Many, if not most of them, are seriously inadequate by Western standards. Should not they be shut down at once? This extreme measure is unnecessary and would cause widespread hardship and serious disruption of the economy. The Chernobyl disaster occurred as a result not only of design faults but also of a whole series of gross operating errors, including switching off the safety devices. With responsible operating procedures, the reactors can be operated safely. Nevertheless, it remains urgent to modify them to bring them up to international standards, and this is now being done with the help of the International Atomic Energy Agency.

4

DESTRUCTION OF THE LANDSCAPE

4.1 Visual Impact

The landscape usually means nature in its original state, untouched by humans. This is often of great beauty, and we want to do all we can to preserve it. Over most of the earth, however, mankind has already left its mark, and it is not always destructive. In many countries the work of farmers and builders over the years has created a man-made landscape that also has great beauty. An example is the Cotswold hills north-west of Oxford, in England, with its gently rolling hills, its well-tended farms, its flocks of sheep and its golden-stone villages and magnificent parish churches.

We want to prevent this from being spoiled by ugly buildings, factories, electric pylons and motorways, so first of all we will consider the visual impact of energy generation.

All power stations are large buildings that have a great visual impact on the landscape. Coal power stations are usually surrounded by huge piles of coal waiting to be burned and by heaps of discarded ash. Oil and nuclear power stations are somewhat smaller, but all have huge cooling towers to reduce the temperature of steam after passing through the turbines. It is possible, to some extent, to locate such power stations in remote areas so as to reduce their visual impact, but, in general, they are blots on the landscape. The main consolation is that each of them produces such a large amount of power that a rather small number suffices for the needs of a whole country.

Oil refineries are hardly beautiful, except perhaps at night, but, fortunately, rather few are required, and oil rigs are far out at sea. A much more serious threat to the environment from oil is the possibility of a tanker running aground, breaking up and spilling thousands of tons of oil over the sea. This oil can travel large distances and pollutes beaches and kills fishes, seals and birds. It takes a long time, and is extremely expensive, to clear up this oil pollution, and the effect on the environment and wildlife is severe and sometimes irreparable. There have been many such environmental disasters from the *Torrey Canyon* on, and hardly a years goes by without another.

Electricity pylons are now a familiar feature of the landscape, and are inevitably associated with power generation. It is possible, but unfortunately prohibitively expensive, to bury the power lines underground. This may, however, at some future time, be overcome by new technologies, such as high-temperature superconductors.

Hydroelectric power is extremely destructive of the landscape. To ensure a continual supply of water, the rivers are dammed to form large lakes, and this usually means that beautiful mountain valleys are destroyed. This is a major tragedy for the people who have lived there for generations, as has happened in many cases, for example in the European Alps. The lake formed by the dam may have some amenity value, but during the year the water level rises and falls, and in the dry season this exposes ugly sterile bands of mud. There is increased public awareness of the destruction of the environment due to hydropower, and environmentalists now make strong protests. A recent example is provided by the dams in Western Tasmania, which threatened to destroy an area of outstanding natural beauty. Energetic protests were only partly successful, but some of the area has been saved.

Hydropower is more dangerous than is usually believed. Dams may look solid, but they can be fractured by earthquakes and undermined by water seepage. If they burst, they can cause not only huge loss of life but also further damage to the environment.

The renewable energy sources have even greater effects on the environment. The energy they tap is very thinly spread, and so inevitably the collectors must occupy a very large area. So far, only wind power has been developed on a moderate scale, in the form of wind farms. To catch the available wind, they must be placed on high ground, where they can be seen for many miles. Many hundreds of large windmills are needed to give the same output as a coal or oil power station, so the wind farms occupy vary large areas. Each windmill requires an access road, and this interferes with agriculture and increases the cost of returning the land to its previous state when the windmill is dismantled. In addition to the visual affect, the rotating blades make a humming noise that is very disturbing to people living nearby. Already it is found that this is so objectionable that people want to move away, and house prices fall.

The other renewable sources, if they are developed, are also likely to have a serious effect on the environment. Tidal power requires a huge barrage across an estuary, such as that of the river Severn, and this changes the whole ecology of the area. The lives and perhaps the very existence of the fishes, the birds and the plants of the area are irrevocably affected.

Wave power requires large structures along the coastline, and these would be visually prominent and disruptive to sailing, swimming and other beach activities.

Some recent figures for the amounts of land in square metres taken up by various forms of power stations for each megawatt of energy generated are: nuclear 630, oil 870, gas 1500, coal 2400, solar 100,000, hydroelectric 265,000 and wind 1,700,000.

4.2 THE GREENHOUSE EFFECT

Another much more radical way that the landscape could be destroyed is by drastic climate change. If the average world temperature varied so that the balance of the ecology was upset and plants and trees died, or if the land itself became flooded, that would indeed be the destruction of the landscape. Plants and trees are already being affected by acid rain, which, because of its effect on health, was discussed in Chapter 3. The more drastic changes could be caused by the greenhouse effect.

Even if we were able to burn pure coal, so that there would be no ash or poisons in the smoke, we would still inevitably produce carbon dioxide. This is the result of burning the coal, so that the carbon in the coal combines with the oxygen in the air to give carbon dioxide. Over the last few decades, it has been found that the amount of carbon dioxide in the atmosphere is steadily increasing. Much of it is removed by plants and trees as they grow, and more is absorbed by the oceans. However, the destruction of forests in so many countries and the large amounts of carbon dioxide produced by power stations and other industrial processes has led to a net increase.

The effect of this increase in the concentration of carbon dioxide in the atmosphere is to retain more of the sun's heat, just like a greenhouse. It might at first be thought that this would be a good thing, we would all be warmer and so need to produce less energy. However, it is not as simple as that. If the earth warms up, part of the polar ice caps will melt, and this raises the sea level, flooding low-lying countries like Bangladesh and Holland. It is very likely that this will also seriously affect the world's weather, with incalculable results.

In addition to carbon dioxide, there are several other gases that contribute to the greenhouse effect, in particular methane, nitrous oxide and the chlorofluorocarbons (CFCs). The last two of these are far more damaging per molecule than carbon dioxide, by factors of about 200 and 4000, but the quantity of carbon dioxide is so much greater that it accounts for more than 60 per cent of the greenhouse effect. The very great damage caused by the CFCs had led to demands that they be banned as a soon as possible.

There is much discussion about the timescale and the magnitude of global warming and of its effects. The whole question has recently been studied by several hundred scientists meeting under the auspices of the World Meteorological Organisation and the United Nations Environment Programme. While there are many uncertainties, the following figures give the best available estimates.

It is established that the mean surface air temperature has increased by 3–6 °C over the last hundred years, and the five warmest years have all occurred in the 1980s. It is estimated that by the year 2100 the temperature will have increased by another 4 °C if nothing is done to reduce emissions, and by about 2 °C if emissions are controlled. If nothing is done, the rise in temperature is estimated to increase the sea level by 60cm by 2100, or by about 40cm if emissions are controlled.

It is possible that the recent floods in Europe and the United States and the drought in Africa are early manifestations of the greenhouse effect. It is, however, very difficult to be sure that this is due to the greenhouse effect, because the weather fluctuates widely from one year to another. The earth's ecosystem is to some extent self-correcting, so it is also possible that the effects are being masked by other changes. It is likely to be some years before we know whether the greenhouse effect is real or not, but if it turns out to be real, it will be too late to do anything about it. There is thus strong international pressure on all countries to reduce their carbon emissions by pledging a percentage reduction each year. This can be done in several ways, all more or less inconvenient. By far the simplest way is to reduce the reliance on fossil fuels.

5

INDUSTRIAL POLLUTION

5.1 Types of Waste

All methods of energy generation, and indeed all industrial processes, inevitably produce waste. This waste is nearly always dangerous, and sometimes extremely dangerous. With increasing industrialisation this waste has become a serious worldwide problem.

The character of the waste, and the means adopted to dispose of it safely, is very different for the various energy sources. Wood is generally burned on a small scale and pollutes the atmosphere in large cities in the poorer countries. Relatively little ash is produced and this is easily dealt with.

Coal power stations produce huge amounts of ash due to the impurities in the coal. Most of this is expelled into the atmosphere as smoke and poisons the air we breathe. Each coal power station produces over a million tons of ash per year, together with five hundred thousand tons of gypsum and twenty-one thousand tons of sludge. Although this material is not particularly hazardous, its disposal is a substantial operation due to the huge amount of waste.

The carbon dioxide produced by coal and oil power stations is responsible for the greenhouse effect, which was considered in Chapter 4.

By far the greatest danger to the environment from industrial wastes comes from a wide range of manufacturing processes, particularly in the chemical industry. These wastes include poisonous gases emitted into the atmosphere, liquid waste discharged into rivers and seas, and solid waste that is stored on land, often without any secure containment. It has been estimated that the amount of hazardous industrial waste in the United Kingdom alone amounts to between five and ten million tons per year.

Since the beginning of the industrial revolution, such wastes have been poured into the environment, with very little attempt to reduce the dangers. The managers of factories were under pressure to maximise profits and they had little thought for the environment. Often the serious affects on the health of their own workers were not understood. Reducing the volume of hazardous waste and ensuring that it is contained is often very expensive, and even if a company wished to clean up its operations the cost would have to come from the profits.

In recent years there has been growing realisation of the health hazards of industrial waste and strict legal controls are now operating in many countries. Even so, they are not always obeyed and there remains a strong incentive to reduce the expenditure on safety measures to the bare minimum. In many industrial areas, where factories have been operating for many years, the ground is often heavily contaminated with dangerous chemicals produced by factories closed long ago. There have been several cases where houses have been built in such areas, and the health of the people living in them suffered severely before the cause was recognised. The cost of cleaning such land is high.

In recent years there has developed a clandestine trade in unwanted chemical waste between some industrialised and some developing countries. Since the disposal of toxic industrial chemical wastes is now governed by strict regulations, it is often very costly to get rid of unwanted waste. Some developing countries, eager to make money but not too worried about the hazards of toxic waste, are willing to accept such waste and then just pile it up somewhere. Some of these waste dumps have been found, with stacks of corroding drums leaking toxic chemicals. Needless to say, this is totally illegal and very dangerous, and strong efforts are being made to put an end to such practices.

Factories producing chemicals are also responsible for some of the most serious disasters. That at Bhopal in December 1984 led to the emission of methyl isocyanate , which, according to the official 1990 estimate, killed 3,800 people and injured 203,500. This disaster is now almost forgotten, whereas that at Chernobyl, which killed about 30 people and caused about a thousand cancers, is still remembered.

5.2. Nuclear Waste

The main hazard due to the operation of nuclear reactors is from the waste they produce, in particular the fission products that remain in the fuel rods. As the uranium is burned in the nuclear reaction, the fission products accumulate until they prevent the reactor from working. To avoid this, spent fuel rods are continually removed from the reactor and replaced with new ones. The fuel rods go to a reprocessing plant, where the uranium and plutonium are separated and used to make new fuel rods. The fission products are highly radioactive and emit various types of particles and rays. There are alpha-particles, which are stopped by a sheet of paper, beta particles, which are stopped by a thin metal sheet, and gamma rays, which need an inch or two of lead to stop them. These nuclear radiations, as they are called, can damage human tissue, and a very high dose can cause loss of appetite, illness and even death.

It is therefore essential to ensure that the fission products do not escape into the atmosphere or enter the food chain. This is done by first storing them in tanks above ground for several decades until most of their radioactivity has decayed away. Fission products are a mixture of many different types of nuclei. Most of them are unstable, and their decay half-lives (the half-life is the time that a sample takes to decay to half its initial intensity) vary over a very large range from thousands of years to a small fraction of a second. The nuclei with short half-lives decay quickly and the radiations they emit produce a lot of heat. This heat must be continually removed to prevent the liquid boiling. When the liquid containing the fission products has cooled down, it is reduced to solid form and fused to form an insoluble glassy or ceramic substance. This is encased in solid steel cylinders and buried deep underground in a stable rock formation. There is then no chance that the fission products will escape and cause harm. Eventually, over the years, the radioactivity of the fission products will decay until it is similar to that of the surrounding rocks. This is a very well-understood process for dealing safely with nuclear waste.

The amount of radioactive waste from nuclear reactors is not large. Every year, a reactor produces about four cubic metres of high-level waste, 100 cubic metres of intermediate level waste and 530 cubic metres of low-level waste. The total amount of high-level waste produced by the United Kingdom nuclear programme from 1956 to 1986 is less than about 2000 cubic metres, about the same volume as an average house. This is a minute quantity compared with the huge amounts of hazardous chemical wastes produced by the manufacturing industries.

There are other types of radioactive waste that come from a wide range of industrial and medical applications of radioactive isotopes. These are used in wrist watches and smoke alarms, and as tracers in medical diagnosis and medical research. The radioactivity of these wastes is low compared with that of the fission fragments, and they can be safely disposed of by burying them in a surface trench or by encasing them in concrete and sinking them deep in the ocean.

The two major sources of energy for the future, coal and nuclear, thus produce very different types of waste. That from coal power stations goes into the atmosphere as well as being produced in solid form. Nuclear power stations, however, discharge almost no waste into the atmosphere; all the waste is in the radioactive fission fragments. The solid waste from both types of power stations can be safely dealt with, but the atmospheric waste due to coal and other fossil fuel power stations is a most intractable problem with most serious consequences for health and for the environment.

There is widespread public anxiety about the effects of nuclear radiations, particularly concerning the cases of childhood leukaemia near nuclear plants. Seven cases occurred in Seascale, in Cumbria, near the nuclear reprocessing plant at

Sellafield, from 1955 to 1983. This number seemed to be much greater than would be expected by chance and received much publicity. It was, however, very difficult to understand how these cases could be blamed on Sellafield, since the amount of nuclear radiation released is far smaller than the natural background.

To estimate the biological damage that can be caused by a particular dose of radiation we must know the relationship between the two quantities. The difficulty is that the doses that cause measurable damage are hundreds of thousands of times larger that the doses received by people living around nuclear installations. We know about the radiation damage due to the massive doses of several hundred rem received by the victims of Hiroshima and Nagasaki, and in the treatment of certain conditions such as rheumatoid spondylitis. In assessing the radiation hazards of nuclear power, however, we are concerned with doses measured in millionths of a sievert, which produce no observable effects. Can we assume a simple linear relationship between dose and damage, so that the probability of contracting cancer is proportional to the dose? There is little direct evidence for this, and some contrary evidence, but it seems to be the safest assumption to make, and is the one adopted for many radiation standards. It is the basis of calculations made by environmental organisations who want to emphasise the hazards of nuclear power. It is, however, the experience of medical studies of the effects of a wide range of substances that are certainly hazardous in large doses that there is no evidence for a proportionate hazard at extremely small doses. This is hardly surprising, since the body has an innate capacity to repair damage; it is only when the defences of the body are overwhelmed by a massive dose that harm occurs. This has been summarised by the French physiologist Claude Bernard, who remarked that at high enough levels everything is toxic, whereas at low enough levels nothing is toxic.

Nevertheless there remain the cases of leukaemia around Sellafield, five in all, far above the number expected statistically. There are, however, other cases of leukaemia clusters around nuclear plants that were never built, or built only after the cases had occurred. It remains important to see if there is an explanation for the Sellafield cases.

A possible explanation is that they are connected with the influx of many people into a relatively isolated rural community, as occurred in Seascale, near Sellafield. The cases then could be due to some virus effect. This hypothesis can be tested by seeing if the effect occurs in the case of similar population movements involved in the construction in rural areas of factories that have no connection with the nuclear industry. This possibility has been examined for several such cases in the north of Scotland, and indeed an excess of leukaemia cases was found. It still remains to establish the mechanism for this effect.

Another possible mechanism was suggested in 1987 by Gardner, who postulated that the children developed leukaemia as a result of their father's exposure to nuclear radiation. He collected statistics that showed a significant correlation between paternal radiation dose and leukaemic children. This led to several Court cases in which families sought compensation from British Nuclear Fuels, the company operating the plant.

The Gardner hypothesis has such serious implications for the nuclear industry that many further investigations were made. These were on the actual process whereby paternal irradiation could lead to childhood leukaemia, the observations of leukaemia in the children of survivors of the atomic bombing of Japan, and more extensive studies of leukaemia around nuclear plants. The results of these studies have now been summarised by Sir Richard Doll, Dr H.J. Evans and Dr S.C. Darby in *Nature* (367.678.1994). They conclude that the Gardner hypothesis is wrong.

The possibility that nuclear irradiation could cause a gonadal (reproductive cell) mutation leading to childhood leukaemia can be studied using data on genetically-determined leukaemia. The detailed statistical knowledge shows that there may be a recessive mutation that could contribute to a number of observed cases. However, 'it effectively excludes any major contribution from the type of mutation that would be required to account for the appearance of the Sellafield cases in the first generation, namely a dominant mutation with a high degree of penetrance'.

Studies by Neel and colleagues of 'the children of atomic bomb survivors, including more than 1500 born to parents who received a gonadal dose of one sievert or larger, revealed no clearly increased frequency of mutations'. These doses are far higher that those received by the Seascale workers.

Further studies were made of all the leukaemia cases in people under 25 years of age in 1958–90 born after 1958 in Scotland and a part of north Cumbria near the Scottish border, and of all children under 15 born near five nuclear installations in Ontario. They found that 'neither set of results supported the probability of a hazard from the father's occupation'. Several other studies have reached the same conclusion.

Thus, the authors conclude that 'the association between paternal irradiation and leukaemia is largely or wholly a chance finding'. They note that there appears to be 'small but real clusters of leukaemia in young people near Sellafield, and some other explanation for them needs to be sought'.

This highly authoritative study should finally lay to rest the fears of nuclear radiations from plants like Sellafield, but whether it will or not depends largely on the mass media. The presence of leukaemia clusters, and particularly the Gardner hypothesis, has been widely publicised by organisations opposed to nuclear power. This has encouraged families with children suffering from leukaemia to seek compensation, but when the scientific evidence was laid before the court, the judgement inevitably went against them.

It is greatly in the public interest that these matters should be treated as objectively as possible, taking full account of the scientific evidence. This would avoid much unnecessary anxiety, and enable the best decisions to be taken concerning our future energy supplies.

A new way of treating nuclear waste is now under discussion. This is to bombard it with high-energy proton beams that break up the radioactive nuclei into less-active ones. Alternatively, the proton beam is allowed to hit a target of some heavy element, such as uranium; this produces a shower of neutrons that can also be used to disintegrate the waste nuclei. Another possibility is to design a sub-critical reactor using thorium as fuel, which can become critical with extra neutrons generated in the way just mentioned. Such a reactor can be used to burn nuclear waste and to generate electricity in the usual way. Thorium is particularly convenient as it is far more abundant than uranium. The advantage of these methods of treating waste is that the radioactivity is actually destroyed, instead of being put in a safe place. It is, however, likely to be rather more expensive.

6

MEDIA AND POLITICS

6.1 Media

Far more poisonous and damaging than physical waste is the printed, verbal and visual waste produced by the mass media. This waste cannot be contained or buried, nor does it decay away. It spreads and proliferates, creating a climate of opinion in which any objective consideration of energy and environment becomes almost impossible.

There are indeed some science journalists attached to the more reputable papers who do their best to present objective discussions, but they are very much in the minority. Throughout much of the media, the overriding priority is to sell as many copies as possible or to achieve the highest audience ratings. Almost everything is sacrificed to this end. Most people do not much like reading objective scientific discussions; they prefer scandal, horror and sensation, and the media are not slow in supplying it. In addition, there are powerful commercial and political forces that have already prejudged the issues and decided the answers.

Most of the questions concerning the optimum energy sources can only be decided wisely after a careful analysis of all relevant factors, and scientists have an essential role in this debate. It is notorious that their participation is not welcomed, indeed they are kept out. Their contribution might tell against the prevailing political view and even if it does not they are liable to change their minds in the light of new evidence.

Thus, scientists are seldom asked for their views and, if they are, their words are often quoted out of context in a way that supports the media line. Sensational and wholly misleading articles are published and any scientist who tries to correct them is usually ignored. The result of decades of misinformation is that the public opinion on energy sources is far removed from reality.

One of the principal results of the exclusion of scientists from the energy debate is that it is innumerate. Opposing views are supported by rhetoric and emotion and illustrated by error and misrepresentation. Facts and figures are conspicuous by their absence. Since, in most cases, the best course of action can only be determined by comparing numerical estimates of pollution, cost or risk, this essential activity is made impossible.

Governments usually have highly-qualified scientific advisors who are able to give them objective and accurate advice, but politicians with their eyes on the next election are extremely responsive to public opinion. They are pressured to take decisions that are contrary to the best advice. In this way they remain popular, but do incalculable damage. It takes a very strong politician to take the right decision in the face of adverse public opinion.

There are genuine political questions that need widespread public discussion. The economic and technological realities define the area of realistic action, but still leave many options open. We have to balance the competing factors of safety, cost and environmental damage. In practical terms, how much are we prepared to pay for extra safety or to reduce environmental damage? These are real questions to which everyone can make a useful contribution. This debate hardly ever takes place because the technological realities that define the area of realistic debate are not known or accepted.

Environmentalists are not immune from political pressure, and many environmental organisations show the same reluctance to listen to scientists. Very frequently they instinctively support what they call the 'benign renewables', apparently unaware of the damage they do to the environment. They have a vision of a pastoral Arcadia with a few picturesque windmills to supply all our energy needs. There are, however, some encouraging signs that a few environmentalists, when they see a wind farm in operation, are becoming aware of the environmental damage they cause.

Large power stations are rightly the concern of all who care for the environment, but if we are not to return to the Stone Age we cannot do without them. What must be decided is which type of power station is the least damaging to the environment, assuming, of course, that it is also safe, reliable and economic. As we have seen, the principal alternatives are coal and nuclear. Nuclear power is usually anathema to environmentalists, to whom it is the symbol of all that is hateful in our modern technological society, and so they prefer coal power stations, which are far more damaging to the environment.

If the advocates of such policies are questioned it rapidly becomes clear that they have not done their sums. They seem not even to know how to study a question scientifically. At best, they have consulted some politically-motivated sociologist specialising in energy matters, yet are markedly reluctant to seek genuine scientific advice. Not only do they not know, they seem not to want to know, or even to know what knowing means.

6.2 POLITICS

As soon as Fermi achieved a self-sustaining chain reaction with his pile of graphite and uranium, those present knew that an immensely powerful new energy source had been found. The next few years saw the secret development of the atomic bomb and its use to end the war with Japan. In the immediate post-war years, many of those who participated in the atomic bomb work at Los Alamos enthusiastically developed reactors to generate electricity for peaceful uses. Within a few years civil power reactors were operating, first in Britain, and then in the United States and several other countries. The media were full of optimistic visions of the future to be made possible by the new energy source.

In many respects it was indeed a success story. The new nuclear power stations proved themselves to be reliable and safe, and their economical aspects compared reasonably well with those of coal power stations. Nuclear power stations were more costly to build, but their operating costs were lower.

Gradually, however, the mood changed. The accident at Three Mile Island severely shook the public confidence. Fears of unknown radiations and memories of the devastation of Hiroshima and Nagasaki came to the fore, and were fanned by media stories of thousands of people who would die because of the accident. Although devoid of foundation such fears, once released, could not easily be calmed. The nuclear industry had neglected to ensure that people were properly informed about the realities of the nuclear age, and the reports of the Three Mile Island accident showed deficiencies in the design and operation of the reactor.

The reaction against nuclear power was strongly supported by political groups who wanted to reduce the power of the West. They stirred up opposition, making full use of the fears of nuclear radiation and the memories of the atomic bombs. Many politicians urged that legislation be introduced to impose a strict safety review before permission to build a nuclear power stations was granted. Up to a point this is indeed necessary, but eventually the legal requirements became so lengthy and cumbersome that they increased the cost to such an extent that electrical power companies found it was no longer economic to build nuclear power stations. Thus, although by that time the United States nuclear power programme was by far the largest in the world, many nuclear power projects were cancelled and no additional stations were built.

The United States has other sources of energy, in particular immense reserves of coal. It was a pioneer in oil production, but now has to import oil. Thus, it could survive without nuclear power stations, but at the cost of increased pollution. Other countries were not so fortunate. France has no oil and little suitable coal, and realised that it is politically undesirable to rely on imports. So they decided to develop nuclear power, and now about 80 per cent of the electrical power generated in France comes from nuclear power stations. Britain has large reserves of coal and oil, and so the development of nuclear power is not imperative. Many other countries in Western Europe have large nuclear power programmes, and nuclear has now overtaken coal in the generation of electricity in that area. Nuclear power is also popular with the 'tiger economies' of south-east Asia, particularly Korea and Taiwan.

Whether a country develops nuclear power is thus partly a matter of need, depending on its indigenous supplies of coal and oil, and partly a matter of politics. In countries where the alternatives are practicable, those opposed to nuclear power can exert sufficient political influence to slow down the use of nuclear power as, for example, in Britain and the United States, or to stop it altogether, as in Germany. Where this happens, coal or oil power stations are used instead of nuclear, causing much greater pollution. This brings with it increased health hazards, acid rain and more carbon dioxide in the atmosphere. This is the legacy of the anti-nuclear movement.

7

PROBLEMS OF DEVELOPING COUNTIRES

The developing countries are the ones that are hardest hit by the energy crisis. Most of them have no oil or coal, and so they are heavily dependent on imported oil for their energy supplies, their industries and their domestic lighting and heating. If the price of oil rises they are immediately in a very difficult situation. They are usually already heavily in debt and there is very little scope for reducing inessential energy consumption to protect essential services. They have no alternative but to reduce oil imports or plunge more heavily into debt, and often both at once.

The poorer countries often export fruit to pay for their imports, but the terms of world trade often change to their disadvantage. For example, before 1973, Costa Rica had to export 28kg of bananas to pay for one barrel of oil. Ten years later, one barrel of oil cost 420kg of bananas. Already the budgets of many African, Asian and South American countries are overstretched by the need to buy oil for essential purposes. The price of oil fluctuates, but on the whole it tends to increase because the oil-producing countries want to maximise their profits and because the richer countries are using more than their fair share.

In the developing world, wood is still extensively used, and if oil is too expensive the increasing demand for fuel leads to the destruction of the trees and the creation of a wasteland. The countryside for miles around many African towns is being stripped of anything that can burn. As soon as saplings are planted, they are ripped up for fuel. The children, instead of attending school, spend hours searching for sticks and twigs. Animal dung, which should be returned to the soil, is dried and burned as fuel. The productivity of the land inevitably falls, leading to a downward spiral of poverty, malnutrition, famine and death.

Such people cannot be helped by providing power stations, because in the rural areas and often in the towns they are not electrified. They cannot afford the wiring and the electrical equipment. What they need is cheap oil, and world oil prices are likely to come down only if the developed countries use less oil, for example by building nuclear instead of oil power stations. As their economies develop, the poorer countries will be able to afford electrification first in the towns and then in the villages, and at this stage they will need power stations. It is particularly desirable that relatively small power stations be developed.

There are already many large towns and cities in even the poorest countries, and their energy needs cannot be met without severe pollution from burning wood and coal. Some of the largest and fastest-growing cities in the world are in relatively poor countries. The population of Mexico City is expected to reach thirty million, and Bombay and Calcutta twenty million each. Already these cities are heavily polluted, so how can they be given the energy they need without increasing this pollution to an intolerable level?

The present world situation is extremely complicated. Developed countries in Europe and North America generally have many possible energy sources and are rich enough to buy what they need from other countries. They are able to enforce pollution controls and on the whole the state of their environment is improving year by year. Other countries, such as parts of eastern Europe and Russia have severe environmental problems due to decades without effective controls, and problems with ensuring a reliable supply of energy. Many, if not most, of the developing countries have a severe energy shortage and environmental problems ranging from erosion and desertification in rural areas and atmospheric and chemical pollution in the cities. So many people in the developing countries lack sufficient energy that there is an overall world energy shortage that requires urgent attention. Since each country has different needs and resources, it must have its own energy policy, and this must be co-ordinated with those of neighbouring countries.

The developing countries vary enormously among themselves. Some are small and poor, with little more than subsistence agriculture. Others, such as India, have substantial industries and the expertise to construct and run their own coal, oil and nuclear power stations. Bombay, for example, has had a nuclear power station for some years. But all these countries, to a greater or lesser extent, find it impossible to cope with the problems of the present and to build the industries needed for the future.

What can be done about this? The poorer countries cannot afford to establish the industries that they need to earn the money to buy oil. If they are fortunate enough to have their own coal they are able to supply their energy needs, but at the cost of pollution. Nuclear power is certainly able to supply the energy needed by the large cities, without adding to the pollution. It would indeed reduce the pollution because if cheap electricity were available it would no longer be necessary to use wood and dung as fuel, which are the principal sources of pollution.

Unfortunately, few of the developing countries can afford to buy nuclear power stations, so some means of providing them cheaply must be found. If they are lent the money needed they will run up huge debts that they will have little hope of repaying, and this leads to the richer countries dominating the poorer ones. They can use their power to force the poorer countries to buy their own goods and accept the prices they set for the material they buy from them. This is economic imperialism.

In a speech to a conference on nuclear power held in Vienna in 1982, the leader of the Vatican delegation emphasised that the development of nuclear power is essential for the poorer countries of the world, and called for its general acceptance as a new source of energy. He said that in order to help the poorer countries a fund should be established to enable them to buy the machines and power stations that they must have if they are ever to escape from their poverty.

There are many serious difficulties, both economic and political. The economic ones could perhaps be overcome by permitting an international company to build a power station in a needy country at its own expense, and then sell the power produced at an economic rate. There would still remain the fear in the company that, after the power station is built, it would be taken over by the Government of the country. If this is done, the reactor could be used to produce fissile material for weapons; this is the problem of nuclear proliferation. This could perhaps be prevented by regarding the land occupied by the power station as United Nations land, and guarding it with UN troops. Given the desperate need for energy in the poorer countries, it should be possible to overcome the economic and political problems of providing it for them.

Some of the most serious pollution occurs in the developing countries, and with it health hazards and the destruction of the environment. The most heavily polluted cities are to be found in these countries. If a country is in a desperate economic situation, and its industries are struggling for survival, they have neither the will nor the resources to worry about safety. Since pollution is no respecter of frontiers, this is a problem for the whole world. Inevitably, any attempt to force pollution regulations on a developing country is liable to run into strong opposition. It may easily happen that the installation of cleansing equipment to reduce pollution, or the adoption of new safety devices, would be so costly that the company could not survive. It is then easy for the developing country to say to the rich industrialised nations that it is all very well for them to adopt a high moral tone about pollution and safety, because they know that it would put them at a commercial advantage relative to the industries in the developing countries. The richer countries developed their industries without worrying about pollution, so why should the developing countries not do the same? This is certainly a serious problem that can only be tackled at the highest international level. It is right to require high standards of pollution control and safety, but this cannot be at the price of bankrupting the industry concerned. This could be overcome by subsidies designed to offset the losses incurred by following the international regulations.

These problems are so complex and inter-related that it is extremely difficult to see how they can be tackled successfully. It is well known that countries hardly ever act altruistically, especially when their vital interests are at stake. It is no good just exhorting either the richer countries to reduce their living standards and their high levels of consumption or the poorer countries to avoid polluting the atmosphere. The force of public opinion puts severe constraints on the decisions of politicians, and in most countries it is not electorally popular to increase foreign aid beyond a token amount. It is practically impossible for governments to impose pollution controls on their own industries that will have the effect of putting them at a severe disadvantage compared with similar industries elsewhere.

To have any chance of success, these problems must be tackled at the highest international level, that is by the United Nations. Only such a body can impose energy policies and pollution controls that have any chance of being obeyed. If all countries agree to impose the same controls, then their industries will be affected in a similar way so that no one gains an advantage over the others. Even then it may be easier to follow the new rules in one country than in another because, for example, the indigenous coal in one country has fewer impurities than that in another. To allow for such differences it may be necessary to arrange appropriate subsidies. Already there is strong pressure to reduce the emission of carbon dioxide to agreed levels. This could be enforced by imposing a carbon tax payable by every county according to the level of emission. This could be extended to form part of a wider network of pollution controls. The income from these taxes could be used to assist the poorer countries to modify their factories and power stations to reduce the level of pollution.

The richer countries could also tax luxury activities that add to pollution and, in this way, set an example that would make it easier for the poorer countries to accept the necessary controls. All such measures will inevitably be highly unpopular and it will not be easy to agree on the necessary regulations, but it is in the long-term interest of every country to take the necessary measures to preserve the earth for future generations.

At the same time as these high-level negotiations it is essential to do all that is possible to educate the peoples of all countries about the need to moderate their lifestyles, reduce pollution and in general to care for the earth. Already it is evident that there is a deep concern for the environment, especially among young people. The phenomenal support for the environmental movements shows the strength of this concern, although, sadly, the activities of some of these movements are counter-productive. As always, effective action requires not only concern and enthusiasm but also awareness of the scientific and technical data that are the essential basis of any realistic policy. As people realise the need, they will more readily support environmental policy decisions.

8

THE THREAT OF NUCLEAR WEAPONS

One of the strongest arguments against nuclear power is that it requires the establishment of a nuclear industry that is able to build not only nuclear power stations but also nuclear bombs. In many respects the technology is very similar. The spent rods of uranium from a reactor can be processed to extract plutonium and this can be made into bombs. Nuclear war is such a terrible threat to the life of mankind that we might even be willing to forego nuclear power if thereby we could remove the possibility of nuclear war.

Whatever we may think about this, it is not an option that is open to us. Nuclear reactors are already operating in many countries, supported in most cases by a nuclear industry. We cannot turn the clock back; nuclear power is here to stay. The danger of nuclear war already exists, and would hardly be affected if we were to demolish all existing reactors. Indeed, it is arguable that such an action would actually increase the danger of nuclear war, as it would have the effect of exacerbating the energy crisis and thus strongly increasing the demand for oil. The resulting international tensions, as the nations scrambled for the remaining oil supplies, could easily be more serious than the threat due to the presently existing reactors.

The technology of nuclear reactors is now well understood and very many nations have enough scientific and technological expertise to enable them to build a nuclear reactor capable of producing weapons-grade plutonium. It is also possible to produce fissile uranium for bombs by separating the two uranium isotopes, using a centrifuge. The manufacture of the bombs themselves is not easy, but the methods used are now widely published. To be an effective weapon it need not be particularly well made. A bomb that explodes prematurely may not be as efficient as those made by an established nuclear power, but it could still cause immense damage.

It has been argued that the operation of reprocessing plants increases the amount of plutonium in the world, and thus the danger of nuclear proliferation. The plutonium can however itself be used to generate nuclear power, particularly in fast reactors, so it is not wasted.

The dangers of nuclear proliferation would be greatly reduced if all reactors were subject to strict international control. This was indeed proposed as long ago as 1946 by Senator Baruch on behalf of the United States, at a time when that country had a monopoly of nuclear weapons. He urged the creation of an International Atomic Development Authority, which would be entrusted with the responsibility for all phases of the development and use of atomic energy from the raw material to the final applications. His plan was that all nuclear activities potentially dangerous to world security would be either owned by the Authority or subjected to effective control, and it would be empowered to control, license and inspect all other nuclear activities. It would also have the duty of fostering the beneficial uses of atomic energy and of keeping at the forefront of research and development of new ideas and techniques. The United States offered to make all of its accumulated experience and equipment available to the Authority. If this far-sighted plan had been accepted, the world would be a far safer place today. Unfortunately the plan was rejected by the Soviet Union, mainly because it seemed to them that it would leave nuclear weapons in the hands of the United States for an unknown period, while at the same time preventing any other nations from making them.

It is now too late for international control in the sense of the Baruch Plan to be a practical possibility, but nevertheless the idea of international inspection is still very relevant. There is always the danger that countries desperately short of energy will construct reactors without taking sufficient care, and this can be prevented if all reactors are subject to inspection and licensing by an impartial international authority.

The danger still remains, and is very real today, that a country will agree to abide by the international regulations and proceed to build up a nuclear industry. It may then suddenly refuse to allow the international inspectors to enter the country, as has indeed happened in Iraq and in North Korea. Immediately, the surrounding nations fear that it has been decided that the hitherto peaceful nuclear industry is about to be used to manufacture bombs.

In the early years of the nuclear arms race, nuclear weapons were frequently tested, and many of the tests released large quantities of radioactivity into the atmosphere. The contamination was particularly heavy for the first hydrogen bombs tests at Bikini and Eniwetok, and the Japanese fishing boat Fukuru Maru was so heavily contaminated by fallout that the fishermen suffered severely from radiation sickness. Thus there was strong international protest when the French announced that they would resume testing in the Pacific. In this case, however, the test explosions were underground, minimising the release of radioactivity into the atmosphere.

Another very serious danger at the present time is that some of the states formed by the break-up of the Soviet Union might decide to sell some of their nuclear weapons expertise, or even the weapons themselves, to countries wishing to increase their international standing and power to threaten other nations.

These are very real dangers and there is little apart from diplomacy that can be done to remove them. These problems are, however, now separate from those of energy production, except that, as already mentioned, world energy shortage is a potent source of international tension, which could be one of the reasons why a nation would wish to develop its nuclear weapons capacity.

9

RESPONSIBILITY FOR THE EARTH

9.1 Basic Principles

Basic to the whole question of energy and the environment is our responsibility for the earth. Are we entitled to use the resources of the earth just as we please, or are there some principles we should follow? Does the earth have absolute rights, or are they relative to those of mankind? Should we just use the earth to satisfy our needs? And what counts as a need? May we also use it for our amusement? Should we be content with just picking nuts and berries, and digging up edible roots, or may we kill animals for food, replace forests by farmland, dam rivers, irrigate plains, mine coal and other minerals, sink oil wells and so build a technological society? Is there any limit to this process?

Intimately connected with these questions are those of our responsibilities to each other. Some people live on rich land with plentiful energy supplies while others are crowded together in poverty on poor land. Are the more fortunate justified in living well and then selling what they do not need to poor people who desperately need it but cannot afford to pay for it? The rich become even richer while the poor sink further into debt or die. Do the rich have an obligation to share their wealth and, if so, how far does this obligation extend? Just to their own families, or to people in the same town or country, or to everyone on earth? And should we not think not only about those living now, but also future generations?

When they tackle their energy problems, countries cannot act alone. Often they need essential fuel from other countries, and the way they generate their energy affects other countries not only indirectly by its economic repercussions but directly by pollution, which is not constrained by frontiers.

Many people believe, or act as if they believe, that all economic matters should be decided by market forces alone. What this means is that greedy and powerful people take far more then they need, leaving the rest in poverty. They do not care if their power plants or factories pollute the rivers and the atmosphere so long as they can maximise their profits.

Most people know that this is simply wrong. The great religions of the world teach that we should respect the natural world. The Judeo-Christian tradition, Islam, Buddhism, Hinduism and Shinto, to mention just a few, all emphasise respect for nature.

Indeed, many early religions made no distinction between mankind and nature. We are surrounded by forces, some evil and some good, over which we have little control. Early peoples personified these forces as gods to be won over with offerings and sacrifices. In such religions, it would have been unthinkable for people to try to control nature and tame it for their own use. In any case, they lacked the knowledge to do this.

The clear distinction between God and nature was first made by the Hebrews about three thousand years ago. They believed in one God, a supreme spirit who alone created the world and keeps it continually in being. Without God the universe would never have existed, and without His continual sustaining power it would immediately lapse into nothing.

The second sharp distinction made by the Hebrews concerns humans. People are a part of nature, completely dependent on God for their being, and yet set apart from nature in an absolutely unique way. Unlike the animals and plants, we have personal freedom. We can decide what to do; we can choose between good and evil. Our special dignity is emphasised in the creation story in the Bible. Humans are created last of all, when the rest of nature has been prepared to receive them, and they are given a special responsibility for the rest of nature. The Psalmist praised the Creator, who made Man 'little less than a god, and has crowned him with glory and splendour, made him lord over the work of his hands, and set all things under his feet'.

9.2 Practical Applications

What does this mean in practice? In primitive societies, areas of the forest are cleared for cultivation and when the soil is exhausted the tribe moves to another site. It is when people settle permanently, with their own land, that the problems arise. Careful farmers soon learn that to ensure continued fertility it is necessary to return to the land what is taken from it, to rotate the crops, to use fertilisers and, depending on the quality of the land, to let it lie fallow periodically.

In our modern technological society we need to apply to the whole earth the same habits of careful husbandry. The problems are, however, immensely more complicated, not only because the economies of distant countries are interlinked but also because industrial developments lead to pollution of the environment. It is no longer just a matter of farmers caring only for their own land; the responsibility for the earth extends also to the whole community at the national and the international levels.

To ensure the care of the earth, regulations governing the use of materials and the control of pollution have been established and enforced. The difficulty is that pollution control, like safety, costs money, and could, if adopted at all stages of an industrial process, easily render it uneconomic. There are thus strong economic pressures to reduce controls to the minimum, or avoid them altogether. However well-intentioned they may be, companies cannot adopt measures that will put them out of business. This difficulty is avoided if national regulations impose the same rules on all companies. In many cases, this applies on an international scale, and so international regulations are necessary. It is not easy to establish the detailed regulations stating exactly what is permitted and what is not, or to ensure that the regulations are obeyed.

It is very easy to say that everything must be made perfectly safe and all pollution must be forbidden. Unfortunately, this is quite impossible. Everything we do carries some element of risk and most industrial processes are polluting to some extent. It is usually relatively easy to make some improvements, but the more that are made, the more difficult it is to make further improvements. Thus, for example, the first half of atmospheric pollution by coal power stations is relatively easy to remove by filters, but the last five per cent is far more costly. So a delicate balance between the cost of the anti-pollution measure and the social cost of the pollution has to be struck before a wise decision can be made.

To achieve this, it is essential to have detailed knowledge of the industrial processes and of the effects of the pollution they produce. Even if this knowledge is available, the pressure of ill-informed public opinion often forces companies and Governments to take what are objectively the wrong decisions. The public appreciation of relative risks is often quite wrong; there is paranoic anxiety about relatively minor risks, while major risks go unrecognised.

An example of this is the very small amounts of radioactivity discharged from nuclear power stations and reprocessing plants, such as that at Sellafield, that constantly feature in the media, while much larger quantities of poisons emitted from coal power stations are hardly ever mentioned. This pressure of public opinion is so strong that Sellafield has been forced to spend £250,000,000 to reduce the discharge of radioactivity by an amount estimated to save one hypothetical statistical life, whereas the expenditure of £5,000 on motoway crash barriers will save one actual statistical life.

9.3 PLANNED OBSOLESCENCE

The two aspects of caring for the earth are the conservation of its resources and the avoidance of waste. To ensure this, manufactured objects should be made to last as long as possible. The manufacturer, however, wants them to wear out as soon as possible, so that we have to buy again. The environmental and commercial interests are thus directly opposed. There are many workmen who take a pride in their work and make things as carefully as possible, with good materials, so that they last a long time. The pressure on manufacturers are, however, so strong that it is now in their interests to make things so that they wear out quickly. This is clearly wrong.

The market is now saturated with some types of electronic equipment. It is very reliable, so the manufacturers have to adopt another method. They announce that they have made a new model and after a certain date they will no longer maintain the old one, or only for an exhorbitant sum. The user is thus forced to buy the new model, even if he or she is perfectly satisfied with the older one.

It is clearly desirable that these practices should be made illegal. Governments, however, are subject to political pressures, and will be told that such laws would curb production and put people out of work. The politicians may even be influenced in other ways by the manufacturers.

CONCLUSION

This survey of the problems of energy and the environment shows that they are part of the much larger problem of the whole future of mankind. Whatever we do, it is inevitable that the population of the world will go on increasing, and the best we can hope for is that it will eventually stabilise at about ten billion by the middle of the next century.

To provide the energy needed by all these people, and to increase the standard of living of the poorer ones, will certainly require a greatly expanded power programme. Since the world's oil supplies will soon reach their maximum, we cannot, without drastic action, hope to keep world energy production even at its present level. To avoid a drop in living standards it is necessary to plan now.

There is a wide range of options. If we do little or nothing, more and more people will find themselves cold and hungry in a few years time. To avoid this, we must make plans for the future and, if they are to have any hope of success, they must be based on accurate knowledge of the capacity of each of the various power sources, its practicability, its cost, its reliability, its safety and its effects on the environment.

Once we have this knowledge, we can think about the sort of world we want to live in, and make our plans accordingly. We must use existing power sources, since it takes a long time to develop new ones. Of those that are available, we have already seen that, although other sources are useful on a relatively small scale, only coal and nuclear have the capacity to supply the bulk of our needs.

We could decide to do without nuclear power and increase the production of coal to supply the bulk of our energy needs. We would then have to accept the increased costs in mining deaths, and the pollution of the atmosphere as we mined more and more coal. We could cut down the pollution by installing more equipment to clean the effluent gases, but this would greatly increase the cost. We would have to decide whether we would be willing to pay the high cost of relatively clean air and reduced pollution. The poorer countries would come off rather badly, because few of them have large reserves of coal, and so they would have to import it from other countries. This would be relatively expensive, since coal is not a convenient fuel to transport large distances, especially to countries far from the sea.

Another possibility would be to build wind and solar generators and hope that they will provide enough energy to replace the oil, without the need to increase coal production. This will greatly increase the cost and will be subject to unpredictable breakdowns. Since wind and solar are relatively hazardous and also have severe effects on the environment, this must be added to the cost of this option. The high cost would make it difficult to provide aid to the poorer countries. Not only would the developed countries be poorer themselves and so have less to spare, but whatever they could afford would provide less energy than the cheaper sources. Reliance on wind and solar power condemns the poorer countries to be without the energy they need.

These two options, or combinations of the two, would avoid the use of nuclear power. It is unlikely that people would be happy with them, once they understand the full implications.

Nuclear power is thus inevitably an important contributor to our future energy needs. Since it already provides almost half the electricity of several West European countries, there is no doubt that it can deliver energy on the scale needed. Hundreds of nuclear power stations have been operating safely for tens of years, so it is a thoroughly well-tested power source. The possible hazards are well understood as are the methods of dealing with nuclear waste. This new technology can provide the bulk of the world energy needs more safely and economically than the alternatives and with less damage to the environment.

Whatever course is taken, the richer countries will manage somehow or another, though probably not without economic crises. Some countries, such as the United Kingdom and the United States, have large supplies of coal and oil, and so need only develop nuclear power in the context of a balanced energy programme. Other countries, such as Japan and France, are not so fortunate and so if they are not to depend on imported fuel they have no alternative but to rely increasingly on nuclear power. Countries that do this will soon enjoy the economic advantages of cheaper electric power.

Through the ages mankind has used in succession wood, coal and oil. As each new energy source has become available more sophisticated uses have been found for the earlier fuels, so that it becomes wasteful to burn them. We are now facing another crisis of transition, as falling oil supplies fail to meet the needs of a rising population. We have found a new source of energy in the nucleus of the atom, and this is not only able to provide the energy we need but is cheaper, safer and cleaner that coal and oil. This new energy comes from uranium, which is otherwise practically useless, and by obtaining energy in this way we conserve the wood, coal and oil so that they are available to satisfy more specialised needs.

The possibility of further nuclear accidents must be assessed as objectively as possible. The science of risk analysis is still in its infancy, but it is infinitely preferable to allowing our future energy policy to remain at the mercy of political decisions. Whatever we do, we cannot entirely eliminate the possibility of an unexpected event or of human negligence and folly. Despite all our essential efforts to improve safety, there will always be accidents and disasters: ships will sink, dams burst, oil rigs capsize or catch fire and mining galleries collapse. It is one of the inescapable conditions of living in a technological society. We must do all we can to minimise accidents, but we must also learn to live with them when they occur. The only alternative is to turn our back on technology and then we would rather rapidly sink back to a primitive level of existence.

The solution to these problems is not made any easier by the political climate in many countries. The result of a long campaign against nuclear power, reinforced by media more concerned with sensation than truth (see Chapter 6.1), is that vital decisions are often decided not by objective scientific evidence but by political considerations. Thus, although careful studies showed that it is safe to dispose of low level radioactive wastes in the sea, it is politically unacceptable.

Another example is provided by the stringent rules imposed on the nuclear industry, concerning radioactive emissions, that far exceed those imposed on other industries. This implies that lives would be saved if uniform pollution controls were applied to all industries. The problems facing politicians are severe enough, without them being forced by political pressure to take decisions that they know are opposed to our long-term benefit.

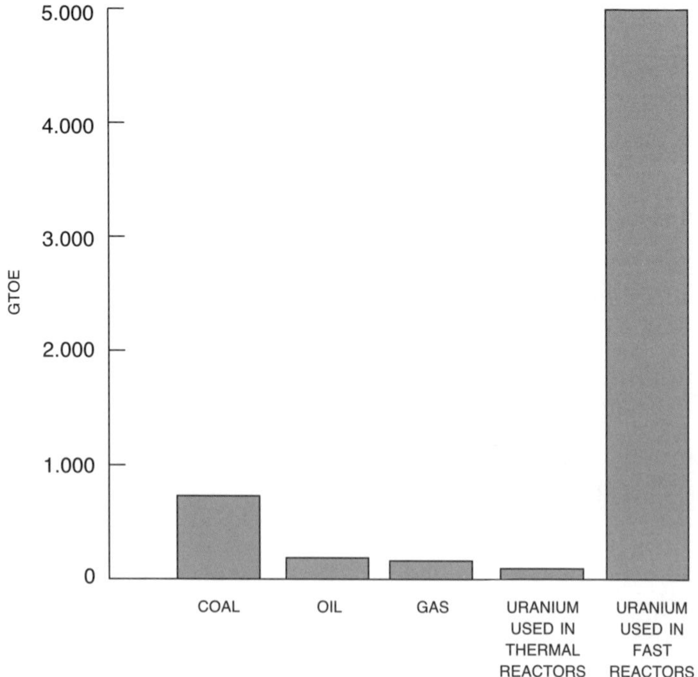

FIGURE 7. *World fuel resources*

It has been very reasonably suggested that a carbon tax should be imposed on all methods of power generation that emit carbon gases into the atmosphere. The rate of tax would be based on estimates of the damage to health and to the environment due to these emissions. This would immediately increase the price of coal power by about a factor of four, making it far less economically acceptable. This shows how much we are really paying for our continued reliance on coal power stations.

Since nuclear power is dependent on uranium, it is important to know how long the supplies will last. In recent years, the discovery of new deposits and the slowing down of the nuclear programme in some countries has led to a large reduction in the price of uranium. As soon as the demand for uranium starts rising the price will go up and this will render economic many mines that were not profitable before. It will be a long time before there is a world shortage of uranium. Before this happens, the fast reactors will have reached the stage of commercial development. It has already been estimated, for example, that in Britain the depleted uranium piling up around nuclear power stations has an energy content greater than that of all the oil in the North Sea. This is illustrated in Figure 7, which shows the world fuel reserves of coal, oil, gas and uranium if used in thermal or in fast reactors. The latter is several times greater than all the rest put together.

There will therefore be enough uranium to last well into the next century, and by then we may hope that the problems of fusion energy will have been solved and fusion power reactors will have become a practical possibility. Scientists and technologists have provided the means of satisfying world energy needs in a way that has little adverse effect on the environment. It is the responsibility of governments and politicians to ensure that the right decisions are taken to bring this about. This will be possible only if they have the support of an informed public opinion. At the present time, ill-informed public opinion often forces governments into adopting policies that they know to be undesirable, thus making it more difficult to provide much needed energy in an economical and environmentally friendly way.

APPENDIX 1
ENERGY UNITS

The basic energy unit is the erg, defined as the work done by a force of 1 dyne moving a distance of 1cm. The dyne is that force which, acting on a mass of 1 g produces an acceleration of 1cm per sec per sec. A joule (J) is 10ergs, and is also defined as the kinetic energy of a mass of 1kg moving at 1 metre per second. Since this is very small for practical purposes, large multiples of the joule are frequently used, particularly the exajoule (EJ) (10^{18} J), the gigajoule (GJ) (10^{9}J) and the megajoule (MJ) (10^{6}J). In the oil industry the energy unit is the tonne of oil equivalent (TOE). 1EJ = 22.7 million TOE, or 1 TOE = 44GJ. Also, 7.3 barrels = 12 tonne. Rates of heat production are measured in watts. A watt is the rate of working of one joule per second. A kilowatt (kW) is 1000 watts, a megawatt (MW) is 10^{6} watts, a gigawatt (GW) is 10^{9} watts and a terawatt (TW) is 10^{12} watts. One kilowatt hour (kWh) is 3.6 MJ. 1EJ per year is 32.2GJ per second or 32.2 gigawatts. One barrel of oil per day is 50 TOE per year. An energy unit used in nuclear physics is the electron volt (eV). One eV is 1.6×10^{-13} J. A million electron volts (MeV) is 1.6×10^{-19} J. Each fission of a uranium nucleus releases 200 MeV = 3.2×10^{-11} J. I gram of uranium 235 undergoing fission releases 82,000 MJ.

APPENDIX 2
RADIATION UNITS

There are many different ways of measuring the intensity of nuclear radiation. Some of those mentioned below are now obsolete, but they are included because they may still be found in earlier publications.

The roentgen is a measure of the ionisation produced in a tissue, and is such that one roentgen produces two billion ion pairs in a cubic centimetre of standard air. The number of atoms in a cubic centimetre is so large that a roentgen ionizes only one atom in ten billion. This unit was originally defined for X-rays and gamma rays, and later a similar unit, the rad, was defined for any ionizing radiation. The rad corresponds to the absorption of a hundred ergs of energy per gram. Since a roentgen delivers about 84 ergs per gram, the two units are very roughly the same. More recently a new unit, the gray, has been introduced. This is defined as the radiation corresponding to an energy absorption of 1 joule per kilogram. Thus 1 gray is equivalent to 100 rad.

Another important unit is the curie, defined as the radioactivity of 1 gram of radium. This may be extended to other radioactive substances by defining a curie as the radioactivity of an amount of that substance that has the same number of disintegrations per second, thirty-seven billion, as a gram of radium. For most purposes this is an inconveniently large unit, so the millicuries and the microcurie, one thousandth and one millionth of a curie respectively, are often used instead. Similar subdivisions are made of the roentgen and the rad. For some purposes it is convenient to use the becquerel, defined as the activity corresponding to one disintegration per second.

Since some types of radiation cause more damage than others, a unit has been defined to provide a standard of comparison between them. This is the relative biological effectiveness (RBE), defined as the dose from 220 keV X-rays causing a specific effect, divided by the dose from the radiation causing the same effect. To make it possible to define the relative effects of various radiations on the human body an effective quality factor, Q, is often used. This has the value unity for X-rays, gamma rays and beta rays, ten for neutrons and protons and twenty for alpha-particles and other multiply-charged particles.

When considering the effects of nuclear radiations on the human body, it is also necessary to include the different sensitivities of the different organs. This is done by defining a unit, the rem. that is the product of the absorbed dose in rads, the effective quality factor and any other modifying factor. The rem can also be defined as the dose given by gamma radiation that transfers 100 ergs of energy to each gram of biological tissue; for any other types of radiation it is the amount that does the same amount of biological damage. A more recent unit, the sievert, has been defined as 100 rem.

FURTHER READING

Eric Ashby, *Reconciling Man with the Environment* (Oxford University Press, 1978).

Eric Ashby and Mary Anderson, *The Politics of Clean Air* (Clarendon Press, Oxford, 1981). The history of the fight for clean air in Britain.

John Blunden and Alan Reddish (Eds), *Energy, Resources and Environment* (Hodder and Stoughton, 1991). A detailed illustrated survey prepared for the Open University.

Edgar Boyes (Ed), *Shaping Tomorrow* (Home Mission Division of the Methodist Church, 1981). A survey of modern technology, with excellent sections on electronics and nuclear power.

Vladimir M.Chernousenko, *Chernobyl: Insight from Inside.* (Springer-Verlag, 1991).

Bernard L.Cohen, *Nuclear Science and Society* (Anchor Books, 1974). A clear and simple introduction to nuclear physics, nuclear power, nuclear radiations and the implications for mankind.

Bernard L.Cohen. *Before It's Too Late: A Scientists' Case for Nuclear Energy* (Plenum Press, 1983). A clear and detailed analysis, with particular attention to comparative risks and media distortions.

Alan Cottrell, *How Safe is Nuclear Energy?* (Heinemann, 1981). A clear and authoritative account of the safety of nuclear reactors.

Clarence Glacken, *Traces on the Rhodian Shore* (University of California Press,1967). An account of the relations of man and nature through the ages.

Wolf Hafele (Ed) *Energy in a Finite World: A Global Systems Analysis* (Ballinger Publishing Company, 1981). An authoritative survey.

Peter Hodgson, *Our Nuclear Future?* (Marshall Pickering, 1982). An account of the energy crisis with a detailed consideration of all the major sources of energy and their associated risks.

John Houghton, *Global Warming: The Complete Briefing.*(Lion Books, 1994). An Authoritative account of the evidence, the implications and the way forward.

Herbert Inhaber, *Risks of Energy Production* (1981). A detailed study of the various types of risk involved in all the major ways of producing energy.

John Passmore, *Man's Responsibility for Nature* (Duckworth, 1980). A philosophical study of beliefs about man's relation to nature.

Alvin M.Weinberg, *The First Nuclear Era: The Life and Times of a Technological Fixer* (American Institute of Physics, 1995). An authoritative account by one of the leaders of the US nuclear power programme.

INDEX